Physics on Your Feet:
Berkeley Graduate Exam Questions

Physics on Your Feet:
Berkeley Graduate Exam Questions

or
Ninety Minutes of Shame
but a PhD for the Rest of Your Life!

Second Edition

Dmitry Budker

University of California, Berkeley, USA and Johannes Gutenberg University, Mainz, Germany

Alexander O. Sushkov

Boston University, Boston, MA, USA

Illustrated by Vasiliki Demas

OXFORD

UNIVERSITY PRESS

OXFORD
UNIVERSITY PRESS

Great Clarendon Street, Oxford, OX2 6DP,
United Kingdom

Oxford University Press is a department of the University of Oxford.
It furthers the University's objective of excellence in research, scholarship,
and education by publishing worldwide. Oxford is a registered trade mark of
Oxford University Press in the UK and in certain other countries

First Edition published in 2015

Impression: 1

Published in the United States of America by Oxford University Press
198 Madison Avenue, New York, NY 10016, United States of America

British Library Cataloguing in Publication Data

Data available

Library of Congress Control Number: 2021934833

ISBN 978–0–19–884236–1 (hbk.)
ISBN 978–0–19–884237–8 (pbk.)

DOI: 10.1093/oso/9780198842361.001.0001

Printed and bound by
CPI Group (UK) Ltd, Croydon, CR0 4YY

To the memory of our teacher and dear friend, Max Zolotorev (1941-2020)

Preface to the Second Edition

In the few years that passed from the publication of the First Edition of the book, we received a lot of positive feedback and encouragement from our readers and colleagues, which has motivated us to continue keeping an eye out for good questions suitable for oral PhD exams, now also at the Johannes Gutenberg University at Mainz and Boston University, the respective schools where the two of us currently teach (and give examinations). Some colleagues have offered us such questions, and, as before, we have come up with a number of "exam-style" questions ourselves.

The Second Edition extends the book rather significantly, with new problems found towards the ends of the sections and new cartoons skillfully drawn by Dr. Vasiliki Demas added throughout the book. We have also taken the opportunity to correct the (surprisingly few) misprints and minor errors noticed in the First Edition.

We hope the readers continue to find Physics on Your Feet useful and perhaps entertaining!

Dmitry Budker
Alex Sushkov

Mainz and Boston
July 2020

Preface to the First Edition

How this book came about

In May 2010, the Physics Department of the University of California at Berkeley where the two of us, at different times, had been Ph.D. students, abandoned the Preliminary Oral Examinations, a.k.a oral prelims, for the first-year graduate students, thus breaking a 60-year-long tradition.

In fact, oral examinations were offered at the Berkeley Physics Department much earlier, however, their most recent format and scope more or less settled by 1950, as described by A. C. Helmholz in his *History of the Physics Department. 1950–1968* (Helmholz, 2004).

The Berkeley prelim was a scary experience for those of us on the "receiving end (A. S.)," and a half-day semi-annual chore for those administering the test (D. B. did this from Fall of 1995 through Spring of 2010; he "missed" taking the oral prelims as he entered the Berkeley Physics graduate program in 1989 as a continuing student, but has had his share of oral examinations elsewhere). Nevertheless, the two of us strongly feel that this has been extremely useful for the students, providing them, perhaps, the first "real-life" scientific-communication experience, and giving an opportunity to look at the beautiful world of physics in some approximation of completeness.

It has also been a truly rewarding experience for the faculty member (D. B.). One learned a lot from the brilliant students, and from the wise colleagues asking truly interesting and profound questions. Some of the material of this book is drawn from the notes taken by D. B. at the exams over the years, as well as from the questions collected by the students and passed as an exam-preparation aid "generation to generation."

The origin of the content of the book, therefore, has collective nature, and we are extremely grateful to the members of the Berkeley physics faculty who have generously allowed us to use their ideas. Unfortunately, many of our sources are no longer alive to ask for permission. We remain in deep gratitude to Profs David Judd, Gil Shapiro, Ronald Ross, and, indirectly, many others.

So what's the point of the book now that the Berkeley orals are no more? We hope this collection will be useful to students (of all ages and everywhere) who wish to refresh and/or test their knowledge of physics, and also to students at the universities that still administer orals. And there are always written prelims, qualifying exams, even at Berkeley (at least, for now) ... The level of the readers we primarily aim at is upper-division undergraduates and first-year graduate students, although some of these problems will certainly be enjoyed by postdocs and distinguished physics faculty, looking for a fun break from or an unexpected contribution to their research.

We have had a lot of fun writing up the problems for this book, and we would like our readers to share this joy (rather than stress out about the upcoming exam, which is unproductive). We are greatly assisted in this by the eye-pleasing drawings prepared by our skillful illustrator, Dr. Vasiliki (Vicky) Demas.

Other books

There are several other collections of problems with the scope and goals partially overlapping with ours. Among these are the following.

- *Scattering and Structures* by (Povh and Rosina, 2005), which is a lovely collection of problems for the German oral Diploma and Ph.D. exams with emphasis on quantum phenomena.
- *A Guide to Physics Problems* by Cahn, Mahan, and Nadgorny (Cahn, Mahan, and Nadgorny, 1994), which is a wonderful collection of written examination problem, also supplied with many delightful cartoons (not to mention the most insightful physics).
- An impressive multi-volume *Problems and Solutions* set by a group of Chinese authors (Zhang, Zhou, Zhang, and Lim, 1995; Lim, 1998; Bai, Guo, Lim, 1991; Lim, 2000) compiled by the Physics Coaching class at the University of Science and Technology in China as a guide for preparation to Ph.D. exams at major American universities. While this collection appears to be very useful, we find the choice of questions and style of solutions to be substantially different from our own.

- *University of Chicago Graduate Problems in Physics, with Solutions* by (Cronin, Greenberg, and Telegdi, 1967) is another great collection, although it generally appears to be more mathematical than this book (and has no cartoons!).
- *University of California, Berkeley, Physics Problems, with Solutions* by (Chen, 1974) is a forty year-old collection of problems based on the Berkeley written Ph.D. exams.

How to use this book

We decided to present the solutions right after the problems, instead of separating them into a different part of the book. Nevertheless, as with all respectable problem books, it is recommended that the reader begins by suppressing the temptation to read or peep into the solution right away, and gives the problem an honest "college try" before consulting with the solution (which may be wrong and/or inelegant, anyway).

Some of the material in the solutions clearly goes beyond of what may be expected at an oral examination. We provide these discussions for those readers who may be interested in more in-depth details about the subject and mark the corresponding passages that can be omitted without sacrificing the quality of exam preparation by placing them in the "aside" environment as this paragraph.

Have fun and good luck!

Dmitry Budker
Alexander O. Sushkov

Berkeley and Harvard
March 2014

Acknowledgements

The original title of this book with which we "lived" for a long time was *Ninety Minutes of Shame (but a PhD in Physics for the rest of your life)*, however, our OUP Editor, Sonke Adlung warned us that this title puts book in serious danger of landing in a wrong section of a bookstore ... We are deeply grateful to Sonke for his patient help and guidance over the years it took to complete this project.

This book would not have been possible without our mentors, colleagues, and students whose ideas inspired many of the problems and solutions found in this book. Their suggestions, guidance, and readings of countless drafts were invaluable, and we sincerely appreciate their contributions.

In the semester preceding the completion of the book (Fall 2013), D. B. taught an undergraduate senior elective course at Berkeley called "Physics for Future Physicist" largely based on the problems in the book. The feedback from the students (several of whom actually non-physicists) was enormously helpful.

Some people we would like to acknowledge specifically are Derek Jackson Kimball, Marcis Auzinsh, Byung-Kyu Park, Sean Carroll, Eugene D. Commins, Mikhail Kozlov, Oleg Sushkov, Victor Acosta, Angom Dilip Kumar Singh, Vladimir G. Zelevinsky, Dmitri D. Ryutov, Richard A. Muller, Nathan Leefer, Ron Walsworth, Max Zolotorev, Steve Lamoreaux, Gregory Falkovich, Pauli Kehayias, Konstantin Tsigutkin, Brian Patton, Michael Solarz, Ron Folman, Szymon Pustelny, Ori Ganor, Michael Hohensee, Mikhail Lukin, Ran Fischer, D. Chris Hovde, Yehuda B. Band, Joel Fajans, Peter Milonni, Sifan Wang, Sean Lourette, Maria Simanovskaia, Simon Rochester, and Tamara Sushkova. Damon English helped to set this project in motion and provided invaluable input at its early stages.

In conjunction with the Second Edition, we would like to acknowledge, in addition to the individuals listed previousley, Arne Wickenbrock, Valerii Zapasskii, Masha Baryakhtar, Lykourgos Bougas, and Diana Saville.

The authors acknowledge the support of their research, that motivated many of the problems in this book.

Contents

1
Mechanics, Heat, and General Physics

Entropy isn't what it used to be.

Physics on Your Feet: Berkeley Graduate Exam Questions: or Ninety Minutes of Shame but a PhD for the Rest of Your Life! Dmitry Budker and Alexander O. Sushkov, Oxford University press. © Dmitry Budker, Alexander O. Sushkov, Vasiliki Demas 2015, 2021. DOI: 10.1093/oso/9780198842361.003.0001

1.1   Bouncing brick

A brick falls flat onto a tennis ball resting on the ground (Fig. 1.1) and bounces back to the height of $h = 1$ m. What height will the ball bounce to? Make reasonable assumptions, for example, neglect the size of the ball compared to the bounce height.[1]

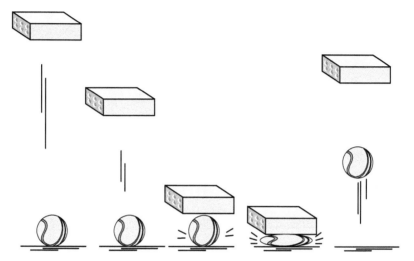

Fig. 1.1 A brick falls onto and bounces off of a tennis ball. As a result, the ball also bounces vertically.

---

[1]This problem was suggested by Prof. G. L. Kotkin.

Solution

When the brick hits the ball, it compresses (squashes) it, and the restoring force, which is mostly due to the air pressure in the ball, is what pushes on the brick first making it to stop, and then turn around and bounce back.

Let us consider the moment in time when the brick is on the way up, and the ball is back to its non-deformed state. Assuming that the ball is at all times in a quasi-equilibrium state, at this moment, the ball is no longer pushing on the brick.

At this moment, the brick begins its free-fall vertical motion with initial velocity v_0 that can be found from

$$\frac{mv_0^2}{2} = mgh. \tag{1.1}$$

Let us now consider the motion of the different parts of the ball. The bottom of the ball is on the ground, and is not moving. On the other hand, the top of the ball is moving at the velocity of the brick v_0 (Fig. 1.1). It is clear, then, that the center of mass of the ball is moving with $v_0/2$. Correspondingly, the ball will bounce to $h/4 = 25$ cm.

Note that tennis players use a related effect to lift a tennis ball at rest on the court the technique involves hitting the ball vertically towards the ground with a racket, upon which the ball bounces up.

1.2 Slippery cone

A climber is trying to climb a slippery mountain with a round conical shape. He has with him a piece of rope with the ends tied together in a knot to form a loop. He throws the loop over the top of the cone and pulls on it to pull himself up, as shown in Fig. 1.2. There is no friction between the rope and the cone surface. If the opening angle of the cone is small (sharp cone), the loop should catch, but if the opening angle of the cone is large (flat cone), the loop should slip off over the top of the cone, as the climber pulls on it.

What is the critical opening angle of the cone so that the rope loop just catches?[2]

Fig. 1.2 A climber pulling himself up with a rope.

---

[2]This problem was suggested to us by Evgeny Kashmensky.

Solution

Suppose the climber has thrown the rope over the top of the cone, but has not pulled it tight yet. Let us cut the cone from its vertex to its base along the straight line passing through the knot in the rope loop. We then unroll the cone into a sector of a circle on a plane (drawing a picture is very helpful at this point, see Fig. 1.3). The knot appears as two points on the radii bounding the sector, these points are equidistant from the vertex, which is the center of the circle. The rope joins these two points, tracing out some curved line contained within the sector.

Now, as the climber pulls on the rope, he puts tension on it. When the rope is tight, it traces out the curve of locally minimal length on the surface of the cone, or a *geodesic*. On the surface of the "unrolled" sector, this is just a straight line between the two points corresponding to the location of the knot. If the sector is less than half of a circle, this straight line is entirely within the sector, and therefore the rope catches somewhere on the surface of the cone [Fig. 1.3 (a)]. If the sector is more than half the circle, the straight line is outside the sector, meaning that, as the rope is tightened, it at some point must cross the center of the circle, thus slipping off the top of the cone [Fig. 1.3 (b)].

The critical cone angle corresponds to the case when the "unrolled" cone forms a half-circle [Fig. 1.3 (c)]. To work out what the opening angle of such a cone is, let us denote the radius of the cone base by r and the length of the cone slant by L. The "unrolled" sector has radius L and arc length of $2\pi r$. When the sector is a semi-circle, $\pi L = 2\pi r$, and $r = L/2$. This means that the critical half-cone angle is $\boxed{\theta = 30°.}$

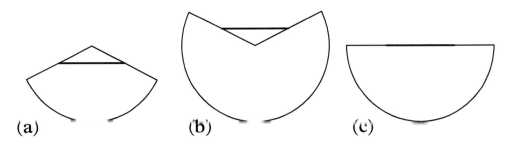

(a)   (b)   (c)

Fig. 1.3 An unrolled cone with the rope on its surface.

1.3 Roach race

Four cockroaches are initially located at the corners of a square with side a as shown in Fig. 1.4. They start moving at the same time with the same speed (not necessarily constant), in such a way that each roach is always moving in the instantaneous direction of its counter-clockwise neighbor. Assume that the size of a roach is negligibly small compared to a.

   What is the distance traveled by a roach until it collides with another?[3]

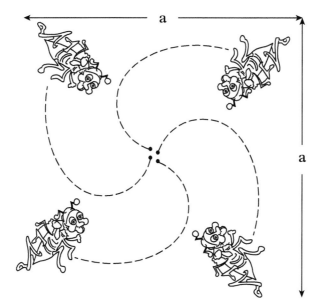

Fig. 1.4 Four roaches start their journey at the corners of a square and each roach moves in the instantaneous direction of its nearest counter-clockwise neighbor.

---

[3]This problem was suggested to us by Elena Zhivun.

Solution

By symmetry, the roaches remain in the corners of a (shrinking and rotating) square at all times till they collide.

Relative-motion problems are often solved most easily by a convenient choice of the reference frame. In this case, such a frame has the origin on one of the roaches, and has an axis pointing in the direction of its neighbor. In this frame, our chosen roach is stationary, while its neighbors approach it along straight lines, covering a distance a till the encounter. Since the motion of the neighbors is always instantaneously perpendicular, one can see that, indeed, the distance covered by a roach till the encounter is a, both in the "convenient" and the stationary frames.

We found that many of our colleagues, somewhat surprisingly, have trouble solving this problem "on the fly."

1.4   Spinning Earth

This problem illustrates some typical and rather instructive estimates that working physicists cover their lunch-time napkins with.

(a) Estimate the number of atoms contained in the Earth.

(b) The angular momentum of an atom or a molecule, if it is nonzero, is on the order of $\hbar$. Estimate the angular momentum of the Earth's rotation per atom in the units of $\hbar$.

Solution

(a) The radius of the Earth is $R \approx 6400\,\mathrm{km} = 6.4 \times 10^8$ cm, so the volume of the Earth is

$$V \approx \frac{4}{3}\pi R^3 \approx 10^{27}\ \mathrm{cm}^3. \tag{1.2}$$

The average density of the Earth is roughly $\rho = 5.5$ g/cm^3, and the most abundant (and typical) element in the Earth is oxygen with an atomic weight of 16. From this, we can roughly estimate the number of atoms in the Earth as

$$\boxed{N \approx \frac{V\rho}{16}N_A \approx 2 \times 10^{50},} \tag{1.3}$$

where $N_A \approx 6 \times 10^{23}$ particles per mole is the *Avogadro number*.

(b) The angular momentum is given by

$$L = I\Omega \approx \frac{2}{5}MR^2\frac{2\pi}{\tau}, \tag{1.4}$$

where $I = 2MR^2/5$ is the *moment of inertia* of a sphere rotating about its diameter, we have approximated the Earth as a uniform sphere, and $\tau \approx 24$ hours is the period of the Earth's rotation. We can now just substitute the mass of the Earth as essentially calculated previously, however, it is better to first divide by the expression of the number of the atoms in the Earth because the mass cancels in the ratio. We have for the angular momentum per atom (in units of $\hbar$):

$$\boxed{\frac{L}{N\hbar} = \frac{\frac{2}{5}MR^2\frac{2\pi}{\tau}}{\frac{M}{16}N_A\hbar} \approx 3 \times 10^{17}.} \tag{1.5}$$

We find that the intrinsic angular momentum of atoms gives negligible contribution to the angular momentum of the Earth, even if all atoms with nonzero internal angular momentum were polarized in the same direction.

1.5   Mechanical oscillator as a force sensor

(**a**) Consider a *damped mechanical oscillator* of mass m, *natural frequency* ω_0, and *quality factor* Q, driven by a force $F(t) = F_0 e^{i\omega t}$ (the force is complex for computational convenience, we take the real part whenever we consider a physical quantity). Calculate the response amplitude x_0 and the phase lag ϕ between the force and the oscillator motion.

(**b**) The oscillator is in equilibrium with a thermal bath at temperature T. Calculate the *root-mean-squared* (r.m.s.) thermal excitation x_T.

(**c**) Suppose the oscillator is cooled to absolute zero. Calculate the r.m.s. excitation x_q of the oscillator due to *quantum zero-point energy*.

(**d**) The oscillator can be used to detect small oscillating forces by tuning its resonance frequency ω_0 close to the force frequency ω and measuring the response. Given the thermal noise calculated in part (b), estimate the force sensitivity F_e of the oscillator after measurement time t. Assume $t \gg Q/\omega_0$.

Solution

(a) The *equation of motion* for the oscillator is

$$m\ddot{x} + \left(\frac{m\omega_0}{Q}\right)\dot{x} + m\omega_0^2 x = F(t), \tag{1.6}$$

where the dots denote time derivatives. Given the form of the driving force $F(t) = F_0 e^{i\omega t}$, let us search for the solution of the form

$$x(t) = x_0 e^{i\omega t}. \tag{1.7}$$

Each time derivative then corresponds to a multiplication by $i\omega$, and Eq. (1.6) gives:

$$x_0 = \frac{F_0/m}{\omega_0^2 - \omega^2 + i\omega\omega_0/Q}. \tag{1.8}$$

Note that $x_0 = |x_0|e^{i\phi}$ is a complex number, its magnitude $|x_0|$ corresponds to the amplitude of the forced oscillations, and its phase ϕ corresponds to the phase between the driving force and the response:

$$\boxed{|x_0| = \frac{F_0/m}{\sqrt{(\omega_0^2 - \omega^2)^2 + (\omega\omega_0/Q)^2}}, \quad \tan\phi = -\frac{\omega\omega_0/Q}{\omega_0^2 - \omega^2}.} \tag{1.9}$$

This steady-state solution is sketched in Fig. 1.5. The frequency width of the oscillator response is ω_0/Q. In the time domain, this can be interpreted as the oscillator *response time* of $\tau \approx Q/\omega_0$ to any sudden change in the force $F(t)$, as can be verified by solving Eq. (1.6) with $F(t) = P_0\delta(t)$, where P_0 is the magnitude of the momentum imparted by the "kick."

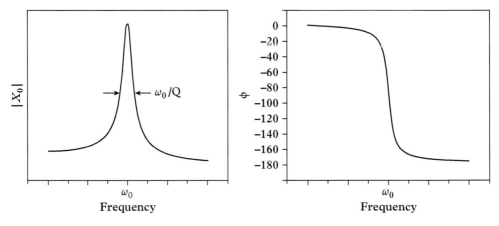

Fig. 1.5 The response of the mechanical oscillator, driven by force $F(t)$. Amplitude response is plotted on the left, and phase response is plotted on the right.

(b) By *equipartition theorem*, the thermal energy of the oscillator in equilibrium with a *thermal bath* at temperature T is given by

$$\frac{1}{2}m\omega_0^2 x_T^2 = \frac{1}{2}k_B T, \tag{1.10}$$

where k_B is the Boltzmann constant. Thus, for a classical oscillator at temperature T,

$$\boxed{x_T = \sqrt{\frac{k_B T}{m\omega_0^2}}.} \tag{1.11}$$

(c) If the oscillator is in its quantum ground state (at absolute zero), its energy is

$$\frac{1}{2}m\omega_0^2 x_q^2 = \frac{1}{2}\hbar\omega_0, \tag{1.12}$$

where $\hbar$ is the Planck constant. Thus, for a quantum oscillator in the ground state,

$$\boxed{x_q = \sqrt{\frac{\hbar}{m\omega_0}}.} \tag{1.13}$$

(d) There are two ways to think about this problem: in frequency domain, and in time domain. In frequency domain, we can approximate the response function in Fig. 1.5 as a rectangular window function of width ω_0/Q, with the thermal noise uniformly distributed over this window. By measuring the oscillator response for time $t > Q/\omega_0$, we are sensitive only to the fraction of this noise in the measurement bandwidth of $\Delta f \approx 1/t$ (neglecting a numerical factor of 2):

$$\frac{1}{2}m\omega_0^2 x_e^2 \approx \frac{1}{2}k_B T \frac{1/t}{\omega_0/Q}, \tag{1.14}$$

where x_e is the oscillator response sensitivity after measurement time t:

$$x_e \approx \sqrt{\frac{k_B T}{m\omega_0^2}\frac{Q}{\omega_0 t}}. \tag{1.15}$$

Alternatively, in time domain, we can consider the thermal noise as a series of random kicks, separated by the oscillator response time of Q/ω_0. The oscillator undergoes a *random walk* with the number of steps $t/(Q/\omega_0)$, and, after measurement time t, the uncertainty in its response is inversely proportional to the square root of the number of steps:

$$x_e \approx \sqrt{\frac{k_B T}{m\omega_0^2}}\frac{1}{\sqrt{t/(Q/\omega_0)}} = \sqrt{\frac{k_B T}{m\omega_0^2}\frac{Q}{\omega_0 t}}, \tag{1.16}$$

which is the same as the result in Eq. (1.15).

Now that we have calculated the measurement sensitivity for oscillator response, all that remains is to convert that to force sensitivity using Eq. (1.9). For optimal sensitivity, the force frequency has to be close to resonance: $\omega \approx \omega_0$, thus

$$x_e \approx \frac{F_e/m}{\omega_0^2/Q},\qquad (1.17)$$

and, using Eq. (1.15), we obtain the force sensitivity after measurement time t:

$$\boxed{F_e \approx \sqrt{\frac{m\omega_0^2 k_B T}{\omega_0 Q t}}.}\qquad (1.18)$$

Note that $k = m\omega_0^2$ is the spring constant, or stiffness, of the oscillator. In order to have a sensitive force sensor, it is necessary to maximize the quality factor Q of the oscillator, and cool it down to low temperature. The quantum limit to the force sensitivity of an oscillator can be obtained from Eq. (1.13).

1.6   Hot-dog physics

(a) When a hot dog is cooked (boiled or fried), its skin often tears. When this happens, the tear is almost always axial (along the length of the hot dog) [Fig. 1.6, left]. Why do they essentially never tear along the circumference [Fig. 1.6, right]?

Fig. 1.6 Hot dogs often tear along the length (left), but essentially never along a circumference (right).

(b) A straight hot dog or sausage, when put on a hot grill, tends to curl into a "C" [Fig. 1.7]. Propose a theory for why this occurs.[4]

Fig. 1.7 When put on a grill, a straight hot dog curls into a "C."

---

[4]This question was suggested by Prof. John Close of the Australian National University.

Solution

(**a**) A hot dog is a long cylindrical shell (casing) made of small intestines of sheep or reconstituted collagen filled with some sort of meat or substitute filling. When the dog is cooked at high temperature, the pressure of the filling rises, which may lead to casing rupture.

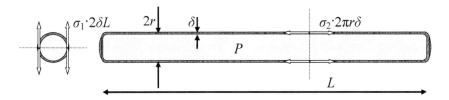

Fig. 1.8 Dimensions of a hot dog and total forces pulling the casing apart in two orthogonal directions, shown schematically with hollow arrows.

Let us assume that the length of the dog is L, the radius of the cylindrical casing is r, and the casing thickness is $\delta \ll r$ (Fig. 1.8). Assuming that the filling is at a uniform pressure P, the force that pulls two halves of the dog apart in a direction perpendicular to the hot-dog axis is $P \cdot 2r \cdot L$ for each half dog. While the dog is intact, this force is balanced by the tension of the casing $2\sigma_1 \delta L$. Here, σ_1 is the force orthogonal to the axis per unit area of the casing. Comparing the two expressions, we get $\sigma_1 = P \cdot r/\delta$.

Let us now look at the balance of forces along the axis. Here, we similarly have $P \cdot \pi r^2 = \sigma_2 \cdot 2\pi r \delta$, which yields for the casing tension in the axial direction: $\sigma_2 = P \cdot r/(2\delta)$, two times smaller than σ_1.

Presumably, the hot dog ruptures when tension reaches the strength limit of the casing material, which clearly happens at lower pressure for rupture parallel to the axis.

By the way, the tension difference also explains why some hot dogs increase their radius when cooked, while growing shorter: the casing expands according to the direction of highest tension, and the length has to shrink if the volume of the "meat" does not change that much.

(**b**) We need to confess here that the search for the solution to this problem led us to experimentation, often accompanied by subsequent consumption of the specimen [Fig. 1.9]. If you have never cooked a hot dog on a frying pan, this is actually a lot of fun. What happens is that the hot dog starts to spontaneously roll around the pan, as if it were alive (a rather dramatic site). Apparently, the reason for this is that bubbles form inside the casing near the hot surface, and there is rapid evaporation into the bubble, causing it to bulge. When this happens, the dog gets a push to roll (which happens in a random direction). This rolling effect confuses the results of our "curling" experiment. To mitigate this, we prevented rolling by slightly holding the dog in Fig. 1.9 while it was cooked on a frying pan (the figure shows the cooked dog removed from the pan and placed on the surface of a cutting board).

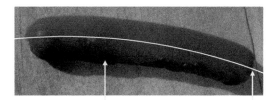

This side was in contact with the
frying-pan surface

Circle approximating the curvature
of the hot dog after heating

Fig. 1.9 A hot dog that was initially straight is seen here to be curled after cooking on a non-stick frying pan. During cooking, the dog was held lightly with a fork to prevent it from rolling (see text).

The result of the experiment shows that, apparently, curling results from the fact that the hotter side of the casing that is next to the hot surface shrinks compared to the opposite side of the casing, thus deforming the hot dog and causing curling. Measuring the radius of curvature of the dog using Fig. 1.9, we estimate the relative shrinkage of the casing close to the hot surface relative to the part of the casing on the opposite side of the hot dog as $r/R \sim 10\%$, where R is the radius of curvature of the cooked dog.

1.7 Ostrich egg

A raw chicken egg, when put into a large pot with boiling water, usually cooks in about five minutes. An ostrich egg has about the same shape as a chicken egg, but its linear size is three times larger. Approximately how long does it take to cook an ostrich egg?

Solution

Let us recall what happens when we put an egg into boiling water. We assume that initially the entire egg is at some uniform "room" temperature. Once in the boiling water, the surface layer of the egg quickly heats up to the temperature of the water (we assume rapid heat exchange in the water outside the egg). Then, due to the *temperature gradient*, heat flows from the outside of the egg towards the yoke, eventually raising its temperature to the point (65°C to 70°C) when it *coagulates*, i.e., becomes solid. At this point, we proclaim the egg cooked.

The heat conduction is governed by the equations

$$\mathbf{q} = -k\boldsymbol{\nabla}T, \qquad (1.19)$$

$$C\frac{\partial T}{\partial t} = -\boldsymbol{\nabla}\cdot\mathbf{q} = k\nabla^2 T, \qquad (1.20)$$

where $\mathbf{q}$ is the heat flux, k is *thermal conductivity*, T is temperature, and C is *specific heat*. The (time dependent) solution of these equations is generally complicated. However, luckily, we do not need to explicitly solve any equations to answer our question.

In fact, the answer can be read off Eq. (1.20). The Laplacian on the right-hand side of the equation contains the second spatial derivative of the temperature. If we change all the spatial dimensions in the problem by some factor, we can restore the original equation if we scale the time (appearing on the left-hand side) by the square of the same factor. Thus, our *scaling argument* tells us that it takes about

$$\boxed{(5 \text{ min}) \cdot 3^2 = 45 \text{ min}} \qquad (1.21)$$

to cook an ostrich egg.

1.8 Joker's pendulum

A pendulum of initial length l_0 is oscillating with initial angular amplitude φ_0 (Fig. 1.10). To play a trick on the observers of the pendulum, Joker slowly pulls the pendulum's thread, such that the relative variation of the pendulum's length over an oscillation period is always small.

How does the oscillation amplitude scale with the length of the pendulum?

Fig. 1.10 The length of a freely oscillating pendulum is slowly changed. We are interested in the corresponding change in the angular amplitude of the oscillation.

Solution

What we have here is *adiabatic evolution* of the pendulum which is characterized by a conserved quantity, the *adiabatic invariant* (Landau and Lifshitz, 1999). We remind the readers that the adiabatic invariant is the volume in the *phase space* (an area in this case of effectively one-dimensional motion), enclosed by the periodic trajectory of the system.

Let us present perhaps a somewhat unexpected mnemonic for remembering what the adiabatic invariant is in the case of a harmonic oscillator: we will appeal of all things to ... quantum mechanics! The energy of a quantum mechanical oscillator is given by

$$E = \hbar\omega(N + 1/2) \approx N\hbar\omega, \tag{1.22}$$

where ω is the oscillator frequency, N is the number of quanta excited, and we neglect zero-point oscillations (this is a classical pendulum, after all!). Now, in quantum mechanics, the adiabatic invariant is the *quantum number* or the label of the state, in this case, N. During adiabatic evolution, the quantum numbers do not change, while the eigenfunctions and eigenenergies themselves may change.

We see that, in this case, the adiabatic invariant is $N = E/\omega$. Since

$$E = mgl\varphi^2/2 \propto l\varphi^2 \tag{1.23}$$

(m is the pendulum's mass; g is the acceleration due to gravity) and $\omega = \sqrt{g/l} \propto l^{-1/2}$, we find that

$$\boxed{\varphi \propto l^{-3/4}.} \tag{1.24}$$

1.9   Slinky magic

A *slinky* is a popular toy that is just a piece of a loose metal or plastic spring. Suppose you hold a slinky in your hand by the top so it stretches and comes to rest (Fig. 1.11). The bottom is not touching the floor. Then you let go of the top.

Describe the initial motion of the bottom end of the slinky after the top is released.

Fig. 1.11 A vertically stretched slinky held by the top (in this case, by Min Ju Lee) and initially at rest is released. The behavior of the bottom of the slinky is often described as "shocking" by the observers.

Solution

The bottom of the slinky remains at rest until the slinky is almost fully compressed, after which the whole thing drops as in free fall. The internet contains many striking slow-motion videos of this.

The way to understand this behavior is to consider the speed with which a perturbation propagates along a stretched slinky. Suppose that, instead of releasing the upper end, one just gives it an abrupt vertical jerk. This action will send a *compression wave* down the slinky, which will propagate relatively slowly, and will reach the bottom with a noticeable delay. In fact, for this delay to be long compared to the characteristic free-fall time over the length of the slinky is a condition for the "slinky magic." This has been nicely explained on a *Radiolab podcast* (`http://www.radiolab.org/blogs/radiolab-blog/2012/sep/10/what-slinky-knows/`), which inspired this problem, by noticing that the bottom "does not know" that the top has been released until the compression waves reach it.

1.10   Lightbulb and coal

A turn of the century politician *Albert Gore* is said to have claimed that a 100 W *incandescent lightbulb* that draws electricity produced by a *coal powerplant* requires about a ton of coal burned to keep it on for a year. Is this correct?[5] How much *carbon dioxide* is released when one burns a ton of coal?

[5]This problem was pointed out to us by Prof. Roger Falcone.

Solution

First, let us figure out how much energy a 100 W lightbulb dissipates if it is on for a year. Incidentally, wattage of a lightbulb refers to the power taken from electricity, not the power actually converted to light.

Using the fact that one year happens to be nearly exactly $\pi \times 10^7$ s, we find that our lightbulb needs $\approx 3 \times 10^9$ J of electrical energy to stay on for a year.

In the next step, we evaluate how much coal would be burned to produce this much electrical energy. We could, of course, google or look up in a handbook that burning a ton of coal releases something like 2.4×10^7 kJ of thermal energy, but this would be cheating (literally so in an oral exam).

Instead, we can crudely estimate the energy by remembering that a typical chemical reaction changes the chemical energy by on the order of 1 eV per atom. Since coal consists of carbon and hydrogen atoms in roughly equal numbers (which means that the mass of coal is dominated by carbon), we have about $(10^6 \text{ g}/13) \times 6 \cdot 10^{23} \approx 4.6 \cdot 10^{28}$ pairs of hydrogen (H) and carbon (C) atoms in a ton of coal. Here we multiplied the number of moles of hydrogen and carbon by the *Avogadro number*. Combining this with the fact that 2.5 oxygen atoms will be needed on average to burn a hydrogen and carbon (you can verify this by remembering that the final products of burning are H_2O and CO_2) so that we have 4.5 atoms changing their chemical state in the reaction, we find the total energy to be about $4.6 \cdot 10^{28} \times 4.5 \times 1.6 \cdot 10^{-19}$ J $\approx 3 \cdot 10^{10}$ J, not too far from the number one finds in a handbook.

Now, the thermal energy produced by burning coal is not all converted to electricity; a typical efficiency of a coal plant is 35%. So, putting this all together, we find that it takes about a third of a ton to keep a 100 W lightbulb on for a year. A much smaller exaggeration than one would have thought given Mr. Gore's reputation!

As for how much CO_2 is released upon burning a ton of coal, we see that for each carbon atom, there are two oxygen atoms in a carbon dioxide molecule of comparable mass (atomic weight 16 for O vs. 12 for C), so it is about three tons of CO_2 that is released.

1.11 Comfortable walking speed

Have you noticed that there is a certain walking speed that feels "comfortable"? If one walks faster, this takes additional effort, and it may be better to switch to running. Walking very slowly also requires some (at the very least, mental) effort.

How does comfortable walking speed scale with the height of a person assuming that the body proportions remain the same? (Incidentally, small children, especially babies, have rather different proportions compared to adults.)

Solution

The key observation here is that when a person is walking comfortably, the extremities (legs and arms) which can be modeled as *physical pendulums* swing at their natural frequency. A person makes two steps during a period of the oscillation of a leg.

Since we assume that the body proportions are the same for people of different height, the length of a step is proportional to the height of the person H. Now, the speed is given by the length of a step times the frequency of the steps, i.e., the number of steps per unit time, which, for a pendulum scales as $(g/L)^{1/2} \propto H^{-1/2}$.

Combining these results, we find that the ratio of comfortable walking speeds scales as $H \cdot H^{-1/2} = H^{1/2}$. To put this into numbers, a person who is two meters tall has a comfortable walking speed that is about 15% faster than that for a person who is 150 cm tall.

1.12   Rotating dumbbell

A symmetric dumbbell is suspended at its center of mass so that it can freely rotate around any axis going through the suspension point. It is made to freely rotate around the vertical axis as shown in Fig. 1.12. This is accomplished by constraining (with a string) the distance between one of the weights forming the dumbbell and the vertical axis. The vertical suspension rod is assumed to be non-deformable.

Describe the motion of the dumbbell after the string is suddenly cut.[6]

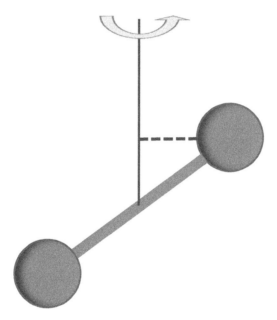

Fig. 1.12 A symmetric dumbbell is suspended at its center of mass and can freely rotate in any direction. Initially, one of the weights is constrained with a string to be at a fixed distance from the vertical axis, and the dumbbell is set to uniformly rotate about this axis.

---

[6]We learned this problem from Prof. Derek Stacey of Oxford.

Solution

Consider the plane, e.g., that of the drawing in Figs. 1.12 and 1.13, which instantaneously contains the centers of both weights and the vertical. The angular-momentum vector **L** of the dumbbell is also contained in this plane and is perpendicular to the dumbbell's axis. In the presence of the constraint, there is a torque that acts on the dumbbell that leads to precession of the angular momentum around the vertical. However, once the constraint is instantaneously removed, the angular momentum immediately ceases its precession, and the dumbbell proceeds to rotate as shown in Fig. 1.13 around the axis collinear with the angular momentum.

One can say that there is no inertia for precession. This is true for both classical and quantum systems such as atomic and nuclear spins. We find that many physicists' intuition fails regarding this, especially when they are put on the spot ...

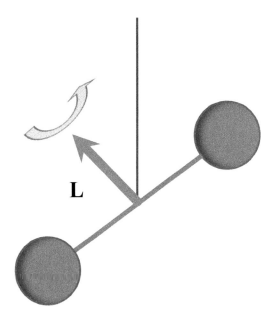

Fig. 1.13 The angular momentum of the dumbbell **L** is perpendicular to the axis of the dumbbell. Once the constraint fixing the distance of one of the weights to the vertical axis is removed, the dumbbell proceeds to rotate in such a way as to conserve **L**.

2
Fluids

If you're not part of the solution,
you're part of the precipitate.

Physics on Your Feet: Berkeley Graduate Exam Questions: or Ninety Minutes of Shame but a PhD for the Rest of Your Life! Dmitry Budker and Alexander O. Sushkov, Oxford University press. © Dmitry Budker, Alexander O. Sushkov, Vasiliki Demas 2015, 2021. DOI: 10.1093/oso/9780198842361.003.0002

2.1 Bubble physics

The physics of *fluids*, quite unfortunately, often "slips between the cracks" in the US university-physics curricula, and students are at times stunned when asked even a simple question about fluids. Here is a fun problem that does not require much prior knowledge of fluids, but does require general knowledge of physics and clear thinking.

(a) A spherical bubble is uniformly moving in a liquid. Describe the motion of the liquid in terms of the *velocity vector field*, i.e., in terms of the magnitude and direction of the liquid flow outside the bubble. Assume that the liquid is *incompressible* and *inviscid* (i.e., there is no friction), that the flow is *irrotational* (zero curl of velocity), and neglect the influence of the container walls.

(b) Based on the result of part (a), calculate the total kinetic energy of the liquid. Neglecting the mass of the gas in the bubble, what is the *effective mass* of the bubble moving in the liquid?

(c) What is the initial acceleration of a bubble released at rest (as in a glass of champagne that you will enjoy celebrating your Ph.D.)?

Solution

(a) We can reasonably assume that the bubble only perturbs the liquid in its vicinity, and that the fluid remains at rest far away from the bubble. Further, since the fluid is incompressible, the total flux into any volume within the fluid must be zero. This means that the divergence of the fluid velocity **v** must vanish:

$$\nabla \cdot \mathbf{v} = 0. \tag{2.1}$$

In addition, we assume that, consistent with our neglecting viscous friction, the curl of the velocity field vanishes as well:

$$\nabla \times \mathbf{v} = 0. \tag{2.2}$$

Now, this should immediately "ring a bell:" we have the same equations for the fluid velocity as we have for static electric and magnetic fields in free space. Therefore, solutions should also be similar.

The one last condition we need in order to determine the velocity field is that the fluid flows around the bubble, i.e., the component of the relative velocity of the fluid with respect to the bubble normal to the bubble's surface is zero.

In electromagnetism, we are used to solutions of problems involving, for example, a *dielectric polarizable sphere* in a uniform external field, in which the field produced by the sphere outside its borders is a *dipole field* (cf. also Prob. 6.6 with magnetic field). In the present case, a direct calculation immediately verifies that the following (educated) guess at the fluid-velocity distribution, indeed, satisfies all our requirements:

$$\mathbf{v}(\mathbf{r}) = \frac{1}{2} \frac{-\mathbf{v}_0 + 3(\mathbf{v}_0 \cdot \hat{\mathbf{r}})\hat{\mathbf{r}}}{(r/R)^3}, \tag{2.3}$$

where $\mathbf{v}_0$ is the bubble's velocity, $\hat{\mathbf{r}}$ is the unit vector along the direction of $\mathbf{r}$, which is the vector from the center of the bubble to the element of the fluid, and R is the bubble's radius.

Notice that the velocities at the "top" and at the "bottom" of the bubble (i.e., at the bubble's "poles") are both equal to $\mathbf{v}_0$, so the fluid at these points, not too surprisingly, moves together with the bubble. In the equatorial plane, the fluid moves in the opposite direction of that of the bubble and with one half the speed (with respect to the bubble, the fluid speed is $3v_0/2$).

Let us comment that nowhere in this solution we used the fact that the spherical object moving in the fluid is actually a bubble. Therefore, within the adopted approximations (such as the absence of *viscosity*), Eq. (2.3) holds also for a solid sphere or a sphere composed of another fluid.

(b) The total kinetic energy of the fluid is

$$K = 2\pi \int_0^\pi d\theta \int_R^\infty r^2 dr \frac{\rho \mathbf{v}(r, \theta)^2}{2}, \tag{2.4}$$

where θ is the polar angle between the bubble's axis and the fluid element, ρ is the fluid density, and the velocity $\mathbf{v}(r, \theta)$ is given by Eq. (2.3).

The integration is straightforward and yields for the total kinetic energy:

$$K = \frac{1}{2} \cdot \frac{4}{3}\pi R^3 \cdot \frac{\rho v_0^2}{2}. \tag{2.5}$$

Let us stress that the kinetic energy associated with the bubble's motion in the fluid, in fact, comes from the motion of the fluid, and not the gas in the bubble (whose contribution to the kinetic energy we neglect), and the effective mass of the bubble is one half the mass of the displaced fluid. This is an example of a common situation in *condensed-matter physics*, where the *effective mass* of an object, for example, an electron or a hole in a semiconductor, could be rather different from its "bare" mass.

(c) The *Archimedes' force* acting on the bubble is directed vertically, and its magnitude is equal to the weight of the displaced fluid. Since the effective mass of the bubble is one half the mass of the displaced fluid, the initial acceleration is $\boxed{2g}$, where $g \approx 9.8$ m/s^2, unless you are drinking your champagne on the moon.

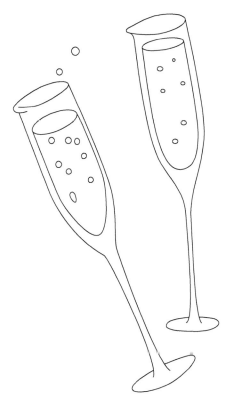

2.2 Bubble and pressure

A wine maker tried very hard to fill a barrel in such a way that, once sealed, there would be no bubbles. She almost succeeded, but a tiny, say, millimeter sized, bubble somehow got trapped near the bottom. Eventually, when the sealed barrel was moved, the bubble let loose and rose from the bottom to the top of the barrel upon the action of the *Archimedes' force*.

Find the change of the hydrostatic pressure at the bottom of the barrel as a result of the bubble's ascent. Assume that wine is an incompressible liquid, neglect evaporation and condensation of the gas in the bubble, neglect surface tension, and assume that the bubble rose slowly enough so that the temperature remained constant. Further, assume that the barrel does not deform, and whatever else you wish to assume (within reason).[1]

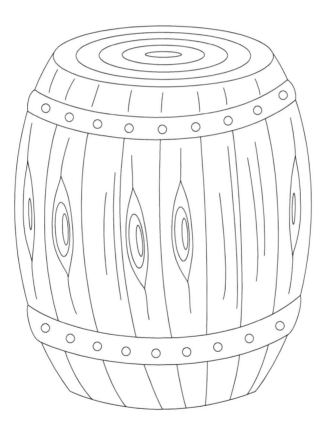

[1]This problem was suggested by Prof. A. S. Shteinberg.

Solution

Let us say that the initial hydrostatic pressure of the gas in the bubble was P_0. When the bubble rises to the top of the barrel, its volume does not change because the liquid is incompressible and the barrel does not deform. Since we assume that the temperature does not change (so the process is *isothermal*) and neither does the amount of gas in the bubble, we are led to conclude, from the *gas laws*, that the pressure of the gas within the bubble remains invariant.

The pressure of the gas is equal to the hydrostatic pressure of the liquid; thus, the bubble's ascent has "transferred" whatever pressure there was at the bottom of the barrel to its top.

Finally, the pressure at the bottom of the barrel is related to the pressure at the top according to

$$P_{bottom} = P_{top} + \rho gh = P_0 + \rho gh, \qquad (2.6)$$

where ρ is the density of the wine, and h is the height of the inner volume of the barrel.

As an example, if the initial pressure was purely hydrostatic $P_0 = \rho gh$, then the rise of our tiny bubble doubles the pressure at the bottom!

Apparently, this effect, that many people find somewhat counterintuitive, is known to cause real problems.

Note that some of the stated assumptions are not necessary (which ones?), while others may not be valid in specific practical situations.

2.3 Holey bucket

A bucket has a hole in its bottom, and so if one fills the bucket with water, it leaks out.

(a) If the level of water in the bucket is H, Fig. 2.1 (left), and neglecting *viscosity*, what is the speed with which the water escapes the bucket? (The bucket is at rest near the surface of the Earth where the gravitational acceleration is g, in case the reader has any doubts ...)

(b) Now suppose that the water escapes from the bottom of a bucket through a length of tubing, see Fig. 2.1 (right), which is sufficiently thin, so the water flow is *viscous flow*, for which the speed is proportional to the hydrostatic pressure at the bottom of the bucket. How is this different from the situation in part (a)?

(c) Suppose a cylindrical bucket, as in part (b), is initially empty, and then, at time $t = 0$, one starts to fill it at a constant rate Γ (for which the units are liters per second, l/s). What is the time dependence of the amount of water in the bucket? Discuss the limiting cases. Note that this very same problem can be formulated for *RC circuits*, populations of energy levels in the course of *optical pumping*, and for myriad other systems.

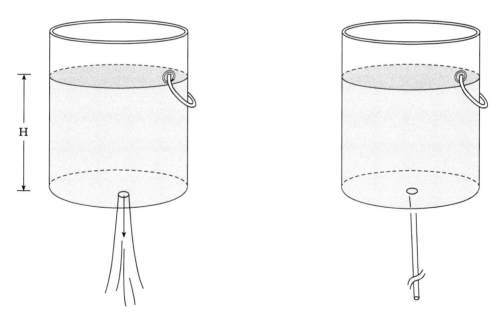

Fig. 2.1 Water leaks from a holey bucket. Left: through a hole at the boom. Right: through a length of tubing.

Solution

(a) There are many different ways to solve this famous problem, for example, to use the *Bernoulli equation*. Let us instead use the energy argument directly. Assuming the hole is sufficiently small, we can neglect the kinetic energy of the water within the volume of the bucket. Suppose the level of the water in the bucket is reduced slightly due to the leak. The energy of the water in most of the bucket has not changed. A thin layer at the top, however, has disappeared. What happened to its potential energy? By energy conservation, it has to go to the kinetic energy of the water escaping the hole. This leads to the expression for the escape speed of

$$v = \sqrt{2gH}, \qquad (2.7)$$

i.e., it is the same as the speed acquired upon free fall from a height H.

(b) Since the hydrostatic pressure at the bottom of the bucket is $\rho g H$, where ρ is the density of water, the rate of the leak scales linearly with H. This is a different scaling from that in part (a).

(c) For a cylindrical bucket, the leak rate is proportional to the amount of water in the bucket P, with a proportionality coefficient that we will call $1/\tau$. We can write a differential equation describing the time evolution of P:

$$\frac{dP(t)}{dt} = \Gamma - P(t)/\tau, \qquad (2.8)$$

which can be directly integrated by rearranging:

$$\frac{dP(t)}{\Gamma\tau - P(t)} = dt/\tau \qquad (2.9)$$

and making a substitution $Y(t) = \Gamma\tau - P(t)$, which yields:

$$\frac{dY(t)}{Y} = -dt/\tau. \qquad (2.10)$$

Integrating this equation, we get

$$Y(t) = Y(0)e^{-t/\tau}, \qquad (2.11)$$

where $Y(0) = \Gamma\tau$ is the initial condition at $t = 0$ corresponding to $P(0) = 0$. Rewriting this in terms of $P(t)$, we get:

$$P(t) = \Gamma\tau\left(1 - e^{-t/\tau}\right). \qquad (2.12)$$

This shows that, initially, the level of water rises linearly with time (as we see by expanding the exponent), but then saturates at the level corresponding to the total equilibrium amount of water in the bucket of $P = \Gamma\tau$. While the equilibrium level depends both on the pump rate (Γ) and the leak size (τ), the characteristic filling-up time is just τ.

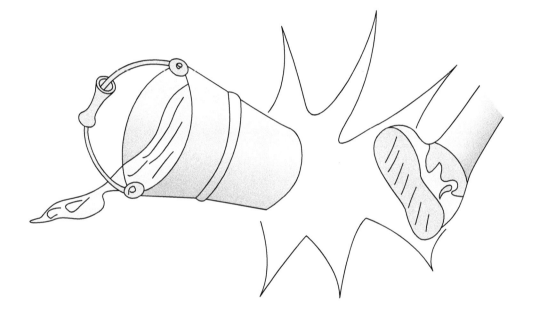

2.4 Surprises in melting and solidification

(**a**) Let us begin with a standard "high-school" question. Suppose there is a cube of ice floating in a glass of water. What will happen to the water level in the glass when the ice melts. Note that the question is of practical importance for those of us who like their glasses full to the brim, and who hate to spill ...

(**b**) If, instead of placing a cube of ice into the water, we place the glass with water into a freezer, and wait until the water partially freezes, we will find that there is a crust of ice on top of the water. In this case, when the ice melts, the water level will, obviously, rise. The question is: how is this consistent with case (a), and can case (b) be considered a limiting case of (a)? Neglect surface-tension effects.

   The facts that an ice cube floats and ice crust forms at the top of the glass are both related to the peculiar property of water, which is shared with just a few other materials, that the density of solid water (ice) is lower than that of liquid. This is crucially important for the existence of life: lakes and rivers freeze from the top, and in many cases do not freeze all the way through to the bottom, preserving living organisms from *death by refrigeration*.

(**c**) Most materials, for example, lead, are denser in the solid phase than they are in the liquid phase. For lead, $(\rho_s - \rho_l)/\rho_l \approx 3\%$, where ρ_s and ρ_l are the solid and liquid densities, respectively. Suppose that we have a tall cylindrical cup full of molten lead. We cool the cup, and let the liquid solidify. The question is: what is the shape of the surface? Assume that the solidification begins on the walls of the cup. Derive the equation for the shape of the surface, i.e., the height of the surface h as a function of radius r. Assume $H \gg R$, where H is the initial height of the liquid, and R is the inner radius of the cup. Again, neglect surface tension, and other unimportant realities (such as finite-height effects) that would distract from the beauty of the result.

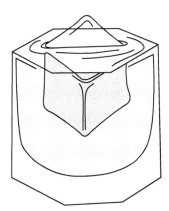

Solution

(a) According to the *Archimedes law*, the floating ice cube displaces the volume of water with weight equal to that of the ice cube. When the cube melts, the produced water exactly "fills in" the displaced volume, and the level of liquid does not change.

(b) The difference with the previous case is that the ice is not floating in the water. In fact, if we were to drill a hole in the bottom of the glass and drain the water, the ice crust would remain stuck in the glass. This is not a limiting case for (a).

(c) This solution follows Siu and Budker (1999). Suppose a molten substance is cooling in a circular cylinder. Assuming that solidification occurs from the side walls of the container inwards in the radial direction, and neglecting the surface-tension effects, we expect the liquid level to drop as a layer of solid is formed because of the higher density of the solid. Consider a newly solidified layer of thickness dr. Let $h(r)$ be the height of solid as a function of radius r. Equating the mass before and after solidification, one obtains a differential equation:

$$\pi r^2 h \rho_l - \pi (r - dr)^2 (h \quad dh)\rho_l + 2\pi r h dr \rho_s. \tag{2.13}$$

Keeping only first-order differentials, we get

$$\frac{dh}{h} = 2\left(\frac{\rho_s - \rho_l}{\rho_l}\right)\frac{dr}{r}. \tag{2.14}$$

With the boundary condition of $h(R) = H$, the solution is:

$$\boxed{h = H\left(\frac{r}{R}\right)^{2\alpha}, \quad \alpha = \frac{\rho_s - \rho_l}{\rho_l} \geq 0.} \tag{2.15}$$

This solution sketched in Fig. 2.2 (top) gives a sharp hole sometimes called *solidification pipe* in the center, the shape of which, for a given container and liquid volume, is determined by α, the fractional density change. An example of a real-life solidification pipe is shown at the bottom of Fig. 2.2.

Note that, for substances that expand upon solidification (water, bismuth, antimony, silicon, and gallium), no "anti-pipe" is formed because the liquid is pushed out by the expanded solidified material and assumes a horizontal level.

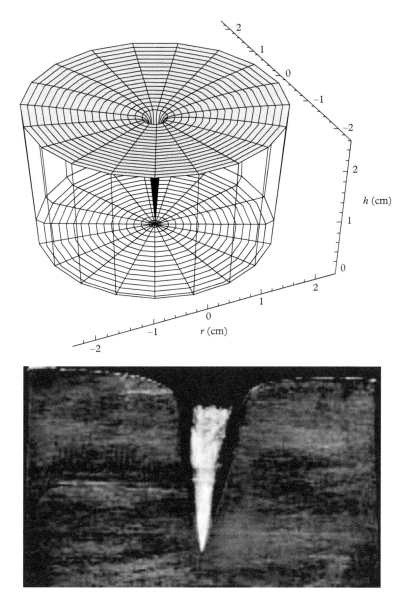

Fig. 2.2 Top: the surface shape predicted by Eq. (2.15) with $H = 2.5$ cm, $R = 2.3$ cm, and $\alpha = 0.025$. Bottom: the actual cross-section of a solder-alloy sample poured into and solidified in a glass beaker. The difference between the observed and predicted shape is largely due to solidification from the top and bottom ignored here.

2.5   Shallow-water and deep-water gravity waves

There are many interesting and important kinds of waves on the surface of water, including *capillary waves* associated with *surface tension*, as well as *gravity waves*, which can be divided into *deep-water gravity waves* when the wavelength of the wave is much smaller that the depth of the water, and *shallow-water gravity waves* corresponding to the opposite limiting case.

(**a**) Neglecting numerical factors (when it is hard to determine them), derive the expression for the frequency of oscillation of a wave with a given wavelength λ. The model sketched in Fig. 2.3 may prove to be helpful. When the movable wall is displaced by a small amount Δx, there appears a difference in the depth of water Δy between the two parts of the tank. When the wall is released, there are oscillations, whose frequency can be estimated. Do this and explain why this model can be used to describe waves beyond those in a weird fish tank.

(**b**) What are the *phase velocity* and *group velocity* of a shallow-water gravity wave?

(**c**) Extend the results of parts (a) and (b) to *deep-water gravity waves*.

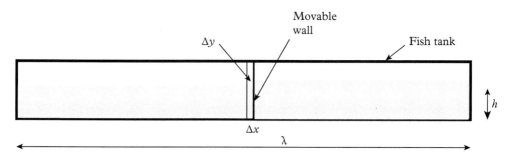

Fig. 2.3 A fish-tank model for estimating the frequency of a shallow-water gravity wave. The massless partition in the middle of the tank can move horizontally without friction.

Solution

(a) If there is a small difference, Δy, in the height of the water on the two sides of the partition, there is a difference in pressure on the two sides of $\Delta p \approx -\rho g \Delta y$, where ρ is the density of water, so that there is a net horizontal force

$$F \approx -\Delta p h b \approx \rho g \Delta y h b, \qquad (2.16)$$

where h is the depth of the water, and b is the width of the tank.

When the partition moves, it actually sets all the water in the tank into (mostly horizontal) motion; therefore, for the purpose of the estimate, we set the *effective mass* of our "*fish-tank oscillator*" to be the entire mass of the water $m = \rho \lambda h b$.

Finally, noting that, due to the conservation of the volume of water, $\Delta y \approx \Delta x h / \lambda$, we can write the *harmonic-oscillator equation* (i.e., the *Newton's Second Law*) for our system:

$$\frac{d^2}{dt^2}(\Delta x) \approx -\frac{\rho g h^2 b}{\rho \lambda^2 h b}\Delta x = -\frac{gh}{\lambda^2}\Delta x, \qquad (2.17)$$

from which we immediately extract the oscillation frequency:

$$\omega_0 \approx \frac{\sqrt{gh}}{\lambda}. \qquad (2.18)$$

We note that nothing about the fish tank "survived" in the final expression, except for the wavelength and depth. This suggests the applicability of our naive model to waves on open waters. Note also that a more careful account for numerical factors for harmonic waves modifies our estimate by a factor of 2π:

$$\boxed{\omega_0 = \frac{2\pi\sqrt{gh}}{\lambda} = \sqrt{gh}k,} \qquad (2.19)$$

where $k = 2\pi/\lambda$ is the *wavevector*.

(b) From Eq. (2.19), we have for the phase velocity:

$$\boxed{v_{ph} = \frac{\omega_0}{k} = \sqrt{gh},} \qquad (2.20)$$

and an identical result for the group velocity

$$\boxed{v_{gr} = \frac{\partial \omega_0}{\partial k} = v_{ph},} \qquad (2.21)$$

since taking the derivative is equivalent to dividing by k for a linear function.

An important observation is that the wave velocities for the shallow waves go as square root of the depth. This explains the peculiar behavior of the water waves as they run into shallow water near a beach. For example, the waves slow down on a shallow part, and thus bend around it.

(c) The deep-water case ($h \gg \lambda$) can be treated in essentially the same way, except that the surface perturbation would not set all the water in the tank into some sort of a motion as we assumed in part (a), but only the water with characteristic depth $\approx \lambda$. So now, instead of Eqs. (2.19–2.21), we have:

$$\omega_0 = \sqrt{gk}, \tag{2.22}$$

$$v_{ph} = \frac{\omega_0}{k} = \sqrt{\frac{g}{k}}, \tag{2.23}$$

$$v_{gr} = \frac{\partial \omega_0}{\partial k} = \frac{1}{2} v_{ph}, \tag{2.24}$$

where, in Eq. (2.22), we once again inserted the correct numerical factor.

An important observation is that the wave velocities for deep-water gravity waves increase with the wavelength. We will encounter this fact in Prob. 2.7.

A nice discussion of surface waves is given by Feynman et al. (1989), Sec. 51—4.

2.6 Tides

Ocean *tides* are a complex phenomenon, where a multitude of factors such as the influence of both Sun and the Moon, the Earth's rotation, continental geography, variations of the ocean depth, etc., all play a noticeable role. In this problem, we will attempt to discuss the tides in a simplified way, yet, retaining the key physical insight into the phenomenon.

Figure 2.4 shows the Earth/Moon/Sun configuration at the time of a *solar eclipse*. Let us assume that we are looking from a direction to where the Earth's rotation axis points (i.e., that of the *North Star, Polaris*).

Which scenario (a), (b), or (c) most closely represents the actual tide pattern? (Assume that the Earth is fully covered by an ocean, which, neglecting the tides, is of uniform depth.)

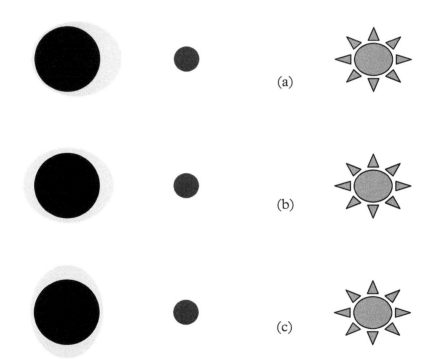

Fig. 2.4 Only one of the three figures (a,b,c) most accurately represents the pattern of high and low tides when the Earth (left), the Moon (center), and the Sun are aligned as during a solar eclipse. Assume that we are looking from the direction where the Earth's rotation axis points. Sizes, distances, and the height of the tides are not to scale.

Solution

Scenario (a) represents a common misconception that, since the Moon and the Sun are pulling harder on the water nearest to them, there will be a tidal bulge on the side of the Earth closest to these bodies. This neglects the fact that there is roughly equal and opposite effective vertical force due to the presence of the Moon and the Sun on either side of the Earth, so high tides occur simultaneously on the opposite sides of the Earth. The consequence of this is the 12 hr (rather than 24 hr) principal periodicity of the tides.

Scenario (b) is what one finds in most books on elementary mechanics. It would have been correct if not for the fast (24 hr period) rotation of the Earth around its axis. This rotation, in fact, leads to the pattern most closely represented by scenario (c).

In order to understand how this comes about, let us, for the moment, forget about the tides, and think of the ocean as a spherical liquid shell covering the Earth. Now suppose that we perturb the spherical shape of the ocean [for example, create wide, small-amplitude bulges on opposite sides of the Earth as shown schematically in Fig. 2.4(b,c) with greatly exaggerated amplitude], and then let the shape freely evolve after that.

What happens then is that the ocean starts "sloshing," which can be thought of as propagation of the shallow (because we assumed wide bulges) *gravity waves* (Prob. 2.5) around the Earth, so, eventually, the bulges will reform when each of the waves goes halfway around. We can thus write the characteristic sloshing period as

$$\tau_0 = \pi R / \sqrt{gh} \approx 25 \text{ hrs}, \qquad (2.25)$$

where we substituted $R \approx 6400$ km for the radius of the Earth and $h = 4$ km for the characteristic depth of the ocean, and used the velocity for shallow waves from Prob. 2.5.

We see that characteristic period of oscillation of our *"ocean oscillator"* is roughly twice longer than the 12 hr tidal perturbation period. We thus have a situation where a resonant oscillator is driven at a frequency ω, considerably exceeding its resonant or eigenfrequency ω_0. But, as is well known to students of elementary mechanics (see also Prob. 1.5, and Fig. 1.5), the driven oscillation in this case is $\approx \pi$ out of phase with the perturbation, thus scenario (c) in Fig. 2.4.

An interesting account of the history of theoretical understanding of the tides, in which the question of this problem played a central role, is given by Darrigol (2009).

2.7 Boat speed limit (hull speed)

The following rule is known amongst seamen: the maximum speed of a *displacement vessel* (e.g., a sailboat), in which part of the hull is submerged in water, is usually well approximated by the following formula:

$$v_{hull} \approx 1.34\sqrt{L_{WL}}, \tag{2.26}$$

where v_{hull} is in *knots* [1 knot = 1 *nautical mile* (1.15 miles) per hour], and L_{WL} is the length in feet of the *waterline* delineating the submerged part of the hull.

The rule of Eq. (2.26) says that one needs a longer boat for higher maximum speed (Fig. 2.5).

Explain the physics that gives rise to Eq. (2.26). Derive Eq. (2.26) from the *dispersion relation* of *deep-water surface gravity waves* [Prob. 2.5(c)]

$$\omega = \sqrt{gk}, \tag{2.27}$$

where ω is the wave frequency, g is acceleration due to gravity, $k = 2\pi/\lambda$, and λ is the wavelength.[2]

Fig. 2.5 The length of this large sailboat *Maltese Falcon* is listed as 289 ft (this is actually not the waterline length, but close enough), and its maximum speed is 19.6 knots, not too far from the prediction of Eq. (2.26). Photo by D. B., who is grateful to Prof. Richard Packard for a great day of sailing on and outside the San Francisco Bay where this photo was taken.

---

[2]This problem was inspired by a discussion with Prof. Jasper Rine of the Molecular and Cell Biology Department at Berkeley.

Solution

The dominant source of resistance for a boat's motion is energy dissipation due to formation of waves, specifically the gravity waves. From the dispersion relation of Eq. (2.27), we derive the expression for the *phase velocity* of such waves:

$$v_{ph} = \frac{\omega}{k} = \sqrt{\frac{g}{k}} = \sqrt{\frac{g\lambda}{2\pi}}, \qquad (2.28)$$

where λ is the wavelength. Let us plug in numbers: with $g = 980$ cm/s^2, we get: $v_{ph}[\text{cm/s}] \approx 12.5\sqrt{\lambda[\text{cm}]}$. We leave it to the reader to verify that this is, in fact, the same as Eq. (2.26) if we identify λ with L_{WL} and v_{hull} with v_{ph}.

What is going on may be described in the following way. As the boat moves through the water, there are various waves formed at the bow and at the stern. The wavelengths of these waves range from very small (slow waves) to the length of the water line (the fastest waves). Imagine now that a boat is moving at the speed of this longest wave. One can argue that the energy transfer to this particular wave is very efficient because the speeds of the boat and the wave are matched. This leads to the formation of large wave crests both at the bow and at the stern. Now imagine that a boat attempts to move faster (for example, by cranking its engines or increasing the sail area). In this case, it will have to start climbing up the "wall" of the forward crest. It does not appear too surprising that a boat moving "nose up" would generate even bigger waves, making it hard to improve the speed.

Interestingly, there are clever ways to avoid the "hull-speed limit," for example, by submerging the vessel to avoid surface-wave formation altogether (submarines), or by getting the boat out of the water, which is known as *hydroplaning*, the technique used by small fast motorboats, as well as racing *catamarans* such as the ones in the America's Cup sailing competition.

2.8 Floating in circles

Observing an object floating on the surface of water (e.g., a *float* used in rod fishing), we discover that when there are waves on the water surface, the elements of water do not move anywhere on average, but instead undergo a periodic motion.

Show that, in the case of deep-water gravity waves (see the previous problems) of small amplitude $A \ll \lambda$, where λ is the wavelength, a surface element undergoes *circular motion*.

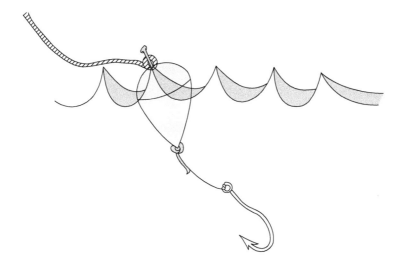

Solution

Let us suppose that the surface shape is described by

$$z = A \sin(kx - \omega t), \tag{2.29}$$

where z is the vertical deviation of the surface from the equilibrium level, k and ω are, respectively, the wave vector and the frequency of the wave, which is assumed to be traveling in the positive $\hat{x}$ direction.

Assuming that the amplitude of the wave is small allows us to first neglect the horizontal displacement of the surface element (think a float marking a particular element), and assume that the vertical displacement δ_z is just the right-hand side of Eq. (2.29).

To find the horizontal displacement, we note that horizontal acceleration of a water element at an inclined surface, neglecting friction, is

$$\ddot{\delta}_x = -g \sin \alpha \approx -g\alpha, \tag{2.30}$$

where g is the acceleration due to gravity, and α is the slope of the surface. Using Eq. (2.29), for example for $x = 0$, we have

$$\alpha = \left. \frac{\partial z}{\partial x} \right|_{x=0} = kA \cos(\omega t). \tag{2.31}$$

Substituting this into Eq. (2.30) and integrating, we get the horizontal speed

$$\dot{\delta}_x = -\frac{kAg}{\omega} \sin(\omega t), \tag{2.32}$$

where we set the integration constant to zero in accordance with the fact that the time averaged velocity must vanish. Integrating once more, we get the horizontal displacement

$$\delta_x = \frac{kAg}{\omega^2} \cos(\omega t) + C, \tag{2.33}$$

where C is the constant of integration.

Finally, using the *dispersion relation* for the deep-water surface gravity waves $\omega^2 = kg$ (see the preceding problems) and comparing Eqs. (2.29) at $x = 0$ and (2.33), we indeed see that a surface water element moves in a circle.

One may observe that, within the approximations that we used, the water element undergoes circular motion with a constant speed $kAg/\omega = A\omega$. However, with respect to the wave, this is not so. Consider the horizontal motion. The water always moves in the $-\hat{x}$ direction in the frame moving with the wave; however, at the crest of the wave, the water moves slower, and at the bottom it moves faster. This is readily observed with a float.

2.9   Boat displacement

A classic problem often presented to students upon their first encounter with the concept of *center of mass* asks to find, ignoring friction, the displacement of a fishing boat that is initially at rest when a fisherman walks from the stern to the bow.

Here, we rise to the next level of sophistication, and pose the same question, but with a more realistic assumption of the friction force acting on the boat being proportional to the boat's velocity.[3]

[3]D. B., then a secondary-school student, learned this problem from Prof. Dmitri Ryutov quite a few years ago.

Solution

In the absence of friction, the boat displacement is such that the center of mass of the boat/fisherman system does not move. In our case, the answer is dramatically different.

Let us assume that the system is initially at rest and that it eventually returns to rest again after the fisherman settles in his new position. Since the system starts at rest and ends at rest, the overall change of its momentum is zero. Now, the change of momentum is the integral over time of the force acting on the system. The only force in the horizontal direction acting on the system externally is that of water friction, which we (realistically) assume to be proportional to the boat's velocity with respect to the stationary water. We are now led to conclude that, since the velocity is proportional to the force, the time integral of velocity is zero, just as it is for the force. Remembering that the time integral of velocity is displacement, we are led to a (shocking?) conclusion that the overall displacement of the boat is, indeed, zero.

2.10   Temperature lapse in the atmosphere

It is observed that the atmospheric temperature typically drops linearly with height at a rate (known as the *temperature lapse rate*) of about 7 K/km up to about 10 km above the Earth surface.

Explain these observations (both the linearity and the order of magnitude of the lapse rate) based on the assumption that the cooling is due to *adiabatic expansion* of air warmed by the surface of the Earth as it rises (or, conversely, that cool air adiabatically heats as it descends).

Solution

Let us assume a certain quantity of gas, for example, one mole. The volume V and pressure P of this parcel will depend on the height z, however, if we assume that the evolution of the parcel of air is *adiabatic*, then the quantity PV^γ is constant, where $\gamma = C_P/C_V$ is the ratio of the *molar heat capacities* of the gas under constant pressure, and constant volume, respectively. From this, it follows that

$$P\rho^{-\gamma} = const, \tag{2.34}$$

where ρ is the density of the gas.

Assuming *ideal gas*, we also have the following relation between the parameters of the gas:

$$P = nk_BT = \frac{\rho N_A}{M}k_BT, \tag{2.35}$$

where n is the *number density* of the gas, k_B is the *Boltzmann constant*, N_A is the *Avogadro number*, M is the molecular weight of the gas (we will assume for the purpose of this problem that air is pure nitrogen with $M = 28$ g/mol), and T is the gas temperature.

Combining Eqs. (2.34) and (2.35), we can exclude ρ and isolate T:

$$T = \frac{M \cdot const^{1/\gamma}}{R}P^{\frac{\gamma-1}{\gamma}}, \tag{2.36}$$

where we substituted N_Ak_B with R, known as the *universal gas constant*.

Having expressed the temperature as a function of pressure, we can now connect the temperature variation with height to the variation of pressure. The latter is found assuming that the gas is in *hydrostatic equilibrium*:

$$\frac{\partial P}{\partial z} = -\rho g, \tag{2.37}$$

where g is the acceleration due to gravity. From this we have:

$$\frac{\partial T}{\partial z} = \frac{\partial T}{\partial P} \cdot \frac{\partial P}{\partial z} = -\frac{M \cdot const^{1/\gamma}}{R}\frac{\gamma-1}{\gamma}P^{-1/\gamma}\rho g = -\frac{M}{R}\frac{\gamma-1}{\gamma}g, \tag{2.38}$$

where in the last step we, once again, eliminated ρ using Eq. (2.34).

Finally, noticing that

$$\frac{\gamma-1}{\gamma} = \frac{C_P - C_V}{C_P} = \frac{R}{C_P}, \tag{2.39}$$

we arrive at

$$\boxed{\frac{\partial T}{\partial z} = -\frac{Mg}{C_P}.} \tag{2.40}$$

We see that, indeed, the adiabatic model predicts linear variation of temperature with height. It should not be taken for granted that any characteristic will change linearly. For example, density drops nearly exponentially with height.

Let us now "plug in the numbers" for the temperature lapse rate:

$$\frac{Mg}{C_P} \approx \frac{28 \text{ g/mol} \cdot 980 \text{ cm/s}^2}{\frac{7}{2} \; 8.31 \cdot 10^7 \text{g cm}^2/\text{s}^2/\text{K mol}} \approx 10^{-4} \text{ K/cm} = \boxed{10 \text{ K/km.}} \tag{2.41}$$

This estimate turns out to be somewhat higher than the experimentally measured value $\partial T/\partial z \approx -7$ K/km, where the presence of water vapor undergoing condensation is a major contributor to the difference.

2.11 Angler's dilemma

An angler fishing in deep waters observes an ocean liner passing by, creating waves heading towards the angler's boat. Fascinated by the site, the angler snaps a picture of the wave pattern on his cell-phone camera. Looking to see how the photo came out, the angler is satisfied with the picture. In fact, he can clearly see a "wavepacket" on the picture (all the time the waves are heading towards him). The angler counts the number of wave crests in the packet seen on the photo and can discern 30 of them.

How many wave crests will the angler experience once the wave packet reaches his boat? Assume that the wavelength is much greater than the size of the angler's boat.[4]

---

[4]This problem was suggested to us by Prof. Dmitri Ryutov.

Solution

The key to the solution is the result discussed in Prob.2.5: the *phase velocity* of a *deep-water gravity wave* is two times higher than the *group velocity*.

What does this mean?

If we see the *wavepacket* moving as a whole, the individual waves are moving twice as fast. In fact, "new" waves appear at the trailing side of the wavepacket, grow in amplitude moving towards the center of the wavepacket, then diminish in amplitude and disappear at the leading edge of the packet. When the angler took a photo of the wavepacket, he sees 30 crests; however, in the time it takes the packet to pass, twice the number of crests, i.e., 60, will be experienced by the angler.

This is a question that very few physicists answer correctly ...

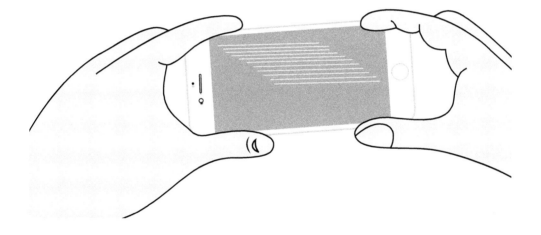

3

Gravitation, Astrophysics, and Cosmology

Have you heard, they opened a restaurant on the moon? Good food, but no atmosphere.

Physics on Your Feet: Berkeley Graduate Exam Questions: or Ninety Minutes of Shame but a PhD for the Rest of Your Life! Dmitry Budker and Alexander O. Sushkov, Oxford University press. © Dmitry Budker, Alexander O. Sushkov, Vasiliki Demas 2015, 2021. DOI: 10.1093/oso/9780198842361.003.0003

3.1 Olber's paradox: why is the sky dark?

Sometimes, an everyday observation can lead to extremely far-reaching conclusions. An example is the so-called *Olber's paradox* based on the observation that the night sky is mostly dark.

(a) Argue that, under the assumptions that the Universe is infinite, homogeneous on large scales, and static, the sky should be infinitely bright.

(b) Give a plausible resolution of the paradox.

(c) Is it possible to resolve the paradox by assuming that the light from distant stars is absorbed by interstellar dust?

Solution

(a) If the Universe were indeed static, the light from a distant star would reach us without changing color, and the intensity of the light would drop as $1/R^2$ with the distance R from the star. Now, imagine a spherical shell centered on Earth with average radius R and a given thickness δR, which we assume to be sufficiently thick, so it contains many stars. The number of stars in the shell, assuming uniform density of stars in the Universe, is proportional to the volume of the shell, $4\pi R^2 \delta R$, and the intensity of the light from the shell reaching the Earth is independent of the radius of the shell. Summing up the shells up to $R = \infty$ (under the assumption of infinite Universe), leaves us with infinitely bright sky – the Olber's paradox.

(b) Two assumptions that lead to the Olber's paradox are not, in fact, correct. First, the Universe is finite. Crudely, the size of the Universe can be estimated by multiplying its age, ≈ 14 Gy, by the speed of light. (In fact, the radius of the observable Universe is usually quoted about three times larger; this is because of the *general relativity (GR)* effect of curved space-time.) Second, the Universe is not static. According to the *Hubble's law*, the stars are receding from us with a speed proportional to the distance R. Because of this, the light from the distant stars is shifted towards longer wavelengths due to the *Doppler effect*. The consequence is a reduction of the intensity of the light reaching the Earth, which is faster than $1/R^2$ without the *red shift*.

(c) An attempt at an explanation of the Olber's paradox by absorption of the light by interstellar dust (while retaining the assumptions of a static infinite Universe) runs into a problem that the dust absorbing intense light from the bright distant sources would heat up, and become a secondary source of radiation, bringing us back to the original problem.

3.2 Gravitational shift of clock rates

Estimate the fractional difference in the rate of identical clocks that are located at different heights (for example, on different floors of the same building with height difference $\Delta h = 10$ m), which arises due to the difference in the gravitational potential of the Earth.

Solution

This is really a *general relativity* (GR) problem, but one only needs to know the most general formulation of the Einstein's *equivalence principle* to solve it.

Consider an experiment in a nonrelativistic spaceship (Fig. 3.1), which is away from sources of gravitational field, but is accelerating at a rate g with respect to an inertial frame. A photon of frequency ω_0 is emitted by a laser near one end of the spaceship, and detected at the other end after traveling a distance h. Now, the time it takes the light pulse to reach the detector is $t = h/c$. However, when the light is detected, the spaceship has accelerated in the direction of light propagation by $\Delta v = gt = gh/c$ compared to when the light was emitted. This means that the light will be detected at slightly different frequency $\omega' = \omega(1 - gh/c^2)$, due to the first-order *Doppler effect*.

Here is where we bring in the great equivalence principle: from the point of view of people in the spaceship, the acceleration should be exactly equivalent to the spaceship sitting still on a launch pad and just feeling the gravitational field with the strength characterized by the same constant g (the gravitational force "down" is equivalent to acceleration "up"). Thus, an experiment on a launch pad should also report a difference in the photon frequency between emission and absorption. But how can this be?

Fig. 3.1 According to Einstein's equivalence principle, it is impossible to tell (without looking out the window) between a uniform acceleration and the action of a gravitational field.

The way out is to conclude that the rate of clocks used to measure the frequency must be different at different heights, so the higher the clock in a gravitational potential, the faster its rate according to

$$\frac{\Delta\omega_{clock}}{\omega_{clock}} = \frac{gh}{c^2}. \tag{3.1}$$

For $h = 10$ m and g being the gravitational acceleration at the surface of the Earth, we have fractional speed-up of the higher clock of about one part in 10^{15}.

In conjunction with this problem, where in the solution we have properly derived the apparent gravitational shift of the frequency for light "climbing" a gravitational potential, just for fun, let us also present a curious but *incorrect* derivation leading to the same result. Here it is.

Einstein has taught us that $E = mc^2$, and we also know that for a photon, $E = \hbar\omega$. Combining these two formulae, we can assign an "effective mass" to a photon of $m = \hbar\omega/c^2$. (Of course, this is exactly the place where we go very wrong: a photon does not couple to gravitational field, and its mass, therefore, is identically zero. But let us carry on for a moment ...) Now, once the effective mass is assigned, we can say that the photon's "kinetic" energy decreases by $mgh = \hbar\omega gh/c^2$ as it climbs up the gravitational potential, which leads us to the same (correct) result for the fractional frequency shift as previously given.

What do we make of the unsettling thought that a blatantly wrong solution has led us to a correct result? Is it pure coincidence?

We would like to propose a point of view that what is going on is really an analysis based on dimensions that is concealed as our incorrect solution. We could have just asked: OK, we do not know exactly what the GR formulae are, but, presumably, the fractional frequency shift should depend on the strength of the gravitational field (g), and how far the photon has traveled (h). Then, we always have the speed of light (c) to make the dimensions right. Putting this together, we immediately arrive at the same result for the fractional frequency shift. Of course, we cannot say anything about numerical factors, but these tend to be of order unity. But, we need luck to get the numerical constant right.

There is another famous example that relates to our discussion having to do with *black holes*. If some mass m is packed in a sphere of some critical radius called the *Schwarzschild radius*, light or matter can no longer escape the object which thus becomes a black hole. The question is, what is the Schwarzschild radius for a given value of m?

A disturbingly incorrect solution would set the non-relativistic expression for escape velocity

$$v_e = \sqrt{\frac{2Gm}{R}}, \tag{3.2}$$

where G is the gravitational constant, equal to c, leading to the correct expression (including the numerical coefficient) for the Schwarzschild radius:

$$R_S = \frac{2Gm}{c^2}. \tag{3.3}$$

Recent state-of-the-art gravitational frequency shift measurements are discussed by Chou et al. (2010).

3.3   Photon fallout

Usual massive bodies, such as, for example, apples, fall due the Earth's gravity with an acceleration of $g \approx 980$ cm/s². What about photons?

You might be wondering: how can one drop a photon? How about this: setup a horizontal flat-mirror *Fabry-Perot interferometer* with high-reflectivity mirrors and let a beam of light bounce back and forth, as shown in Fig. 3.2.

(a) Does a photon fall in the gravitational field? If yes, what is the acceleration?

(b) If your answer to part (a) is affirmative, estimate by how much would a photon fall in a cavity with mirror separation $L = 3$ m and a high-quality low-loss mirror coating that supports 10^6 photon round trips (such high-quality mirrors and coatings are, in fact, possible with modern technology). Briefly discuss what may complicate observation of this effect.

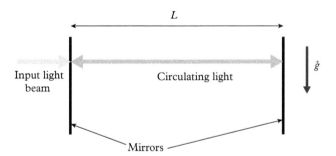

Fig. 3.2 A possible conceptual setup for observing whether photons, in fact, fall in the gravitational field of the Earth.

Solution

(a) A photon is a massless particle, and one might think that, as such, it is not affected by gravity. However, according to the theory of *general relativity (GR)*, the presence of gravitating bodies distorts the space-time, and, to answer the question, we somehow have to figure out what the effect of this distortion is. Luckily, there is a way of doing this without employing any of the mathematical apparatus of GR. The only thing that we need is the Einstein's *equivalence principle*, which says that the effect of gravity is locally equivalent to the effect of a $-\mathbf{g}$ acceleration of the system.

In this case, if we launch the light into a vertically accelerating interferometer, the effect will be exactly as if the photon is falling "like a rock" (or an apple) in the Earth's gravitational field.

(b) The light's round-trip time in the cavity is $2L/c = 2 \cdot 10^{-8}$ s, so a million round trips take $t = 2 \cdot 10^{-2}$ s. The distance that our photons will drop in this time is $gt^2/2 \approx 0.2$ cm. This seems like a large enough distance that opticians might need to worry about photons "falling out" of interferometers!

Of course, there are various effects that prevent such "fallout," including diffraction that tends to blow up the beam size. Moreover, the interferometer mirrors are often made concave, and the resulting *stable-interferometer* configuration tends to compensate the beam displacement. Also, the interferometer needs to be aligned to ensure a proper relative orientation of the mirrors. Typically (if we are not after the falling photons), this is done by maximizing the light confinement time in the cavity. Such a procedure may compensate for the gravitational effect, unless we do something clever, like aligning the mirrors for a vertically oriented interferometer, and then rotating it horizontally. The magnitude of the displacement that we have estimated is rather stunning ...

Finally, let us remark that this effect is closely related to light deflection upon passage in the vicinity of gravitating bodies such as the Sun,[1] which historically was one of the first experimental verifications of general relativity, as well as being the essence of *gravitational lensing* studied in modern astronomy.

---

[1]The reader is cautioned that a naive argument along the lines of what we do here gives one half of the result of a general-relativity calculation for the magnitude of deflection because there is no single local frame where the equivalence principle can be directly applied.

3.4 Planck mass and length scale

What is the physical meaning of the Planck mass and Planck length? Derive their analytical expressions, and obtain approximate numerical values in grams, eV/c^2, and centimeters.

Solution

Consider a point-like particle with mass m. The degree of "quantum delocalization" of such a particle is characterized by the *Compton wavelength*

$$\lambda_\varrho = \frac{\hbar}{mc}, \tag{3.4}$$

This can be understood as follows. If the particle is localized to a spatial extent of λ_c, then the momentum associated with such a localization, from the position-momentum uncertainty relation, is

$$p \sim \frac{\hbar}{\lambda_c} = mc, \tag{3.5}$$

so that the associated relativistic kinetic energy $\sim mc^2$ is on the same scale as the rest energy. Also, λ_c is the distance a particle moving with speed c travels in a time $\hbar/(mc^2)$, the inverse of the angular frequency scale associated with energy mc^2.

The *Planck mass*, m_p, is the mass for which the gravitational self-energy on the corresponding Compton-wavelength scale is equal to the rest energy:

$$\frac{Gm_p^2}{\lambda_c(m_p)} = \frac{Gm_p^3 c}{\hbar} = m_p c^2, \tag{3.6}$$

where G is the universal *gravitational constant*. Solving Eq. (3.6) for m_p yields

$$\boxed{m_p = \sqrt{\frac{\hbar c}{G}}.} \tag{3.7}$$

Substituting the numerical values of the fundamental constants, we obtain

$$\boxed{m_p \approx 2 \cdot 10^{-5}\ \text{g},} \tag{3.8}$$

$$\boxed{m_p c^2 \approx 10^{28}\ \text{eV} = 10^{19}\ \text{GeV}.} \tag{3.9}$$

Finally, substituting the Planck mass into Eq. (3.4), we obtain the *Planck length scale*

$$\boxed{\lambda_p = \frac{\hbar}{m_p c} = \frac{(\hbar G)^{1/2}}{c^{3/2}} \approx 1.6 \cdot 10^{-33}\text{cm}.} \tag{3.10}$$

3.5   Rotation of stars around the center of a galaxy

Measurements of the velocities of stars rotating around the centers of galaxies show that the rotation velocities of stars at a galaxy's periphery are approximately independent of the orbit radius. These velocity distributions, known as *galaxy rotation curves*, are taken as strong evidence of the presence of *dark matter*, which dominates the galaxy's gravitational field, but is not visible to us, presumably, due to the weakness or absence of its interactions via forces other than gravity (i.e., the *electroweak* and *strong forces*).

What would the expected rotation curve of a galaxy be in the absence of dark matter?

Solution

Let us assume that the mass of the galaxy is m, and that a star is moving in a circular orbit at the galaxy's periphery, so that most of the mass of the galaxy is contained within the sphere with the radius R equal to that of the orbit. For a circular orbit, we have

$$\frac{v^2}{R} = \frac{Gm}{R^2},$$ (3.11)

where v is the orbital velocity, and G is the *gravitational constant* (a.k.a. *Big G*), so that

$$v = \sqrt{\frac{Gm}{R}} \propto R^{-1/2}.$$ (3.12)

It follows that the total mass within radius R should be proportional to R in order to explain the observed "flat" rotation curves, so the dark matter extends beyond the visible galactic matter. In fact, from the linear increase of mass with radius, it follows that the density of dark matter falls off as $1/R^2$ in the *galactic dark-matter halo*. It turns out that such density distribution is what one can expect under certain assumption regarding dark matter, for instance, if one assumes that it is a "gas" of interacting (among themselves) objects at thermal equilibrium (see, for example, Peebles 1993).

Understanding the origin and composition of dark matter is among the most important goals of contemporary physics.

3.6 Ultralight dark matter

We know that most matter in the Universe is *dark matter*, but as of the time of this writing, we still do not know what dark matter consists of. Various candidates were proposed including *massive astrophysical compact halo objects (MACHOs)*, *weakly interacting massive particles (WIMPs)*, etc., but none has been conclusively observed so far. In this problem, we discuss a class of dark-matter candidates, the *ultralight dark matter*.

(a) In order to explain the rotation curves of galaxies (Prob. 3.5), in particular, the Milky Way, we need to assume local energy density of the dark matter of $\approx$ $0.4\,\text{GeV}/\text{cm}^3$ (which is about one proton mass per three cm^3).

Assuming dark matter consists of some particles of mass m, how many such particles are there in a cm^3 of volume?

For example, the *CASPEr-Wind-Low Field* experiment will be looking for dark-matter particles with a mass corresponding to a frequency of $\approx 1\,\text{MHz}$. If these particles are indeed dark matter, how many such particles do we expect in a cm^3? How does this compare to the number of nucleons per cm^3 of air in your room?

(b) In order for our dark-matter particles to be confined within a galaxy, they need to move with speeds that are less than the *escape velocity* for the galaxy. This turns out to be typically on the order of $v = 10^{-3}c$, where c is the speed of light. Thinking of de Broglie waves corresponding to our dark matter particles and assuming that they all move with some random velocities on the order of $10^{-3}c$, estimate the *coherence time* and *coherence length* of an ensemble of such waves. In other words, if we have, for example, de Broglie waves corresponding to two particles and they are, say, in phase at a certain point at a certain time; how long will these typically remain in phase at this location? Or, if we look at a given time, how far in space do we need to go to see that the two waves are no longer in phase?

(c) Theorists say that if dark-matter particles are lighter than some m^*, they have to be *bosons* (assuming that only one type of particles contributes to dark matter). Why is that? How do we go about estimating the value of m^*?

Estimate the value of m^*c^2 in eV.

Solution

(a) The number density of the dark-matter particles, under the assumption that all dark matter consists of particles of the same mass m is

$$n_{dm} \approx \frac{0.4 \text{ GeV/cm}^3}{mc^2}. \qquad (3.13)$$

The 1 MHz frequency corresponds to about $4 \cdot 10^{-9}$ eV, so the density of such particles should be $\approx 10^{17}$ cm^{-3}. One cubic centimeter of air contains about $3 \cdot 10^{19}$ molecules, the majority of which are nitrogen with 28 nucleons each, which comes to about 10^{21} nucleons per cm^3, several orders of magnitude more than the previous estimate for 1 MHz dark-matter particles.

(b) The total energy of a nonrelativistic dark-matter particle is dominated by the rest energy mc^2 with an additional correction on the order of mv^2, which comes to about 10^{-6} of the rest energy for $v \approx 10^{-3}c$. This means that the de Broglie waves dephase during roughly 10^6 periods of the oscillation whose frequency corresponds to the energy of mc^2, so that the *coherence time* is

$$\tau_c \approx 10^6 \cdot \left(\frac{mc^2}{2\pi\hbar}\right)^{-1}. \qquad (3.14)$$

For 1 MHz dark-matter particles, this comes to $\tau_c \approx 1$ s.
    Coherence length L_c can be estimated as a product of τ_c and the particle velocity v (we invite the reader to derive this result using the concepts of *phase velocity* and *group velocity* of the de Broglie waves), so that

$$L_c \approx 10^3 \cdot \left(\frac{mc^2}{2\pi\hbar c}\right)^{-1}. \qquad (3.15)$$

For 1 MHz dark matter particles, this comes to $L_c \approx 300$ km.
    Coherence time and length are important parameters for the design of experiments searching for dark matter. For instance, detectors within a coherence length should all detect dark-matter signals that are in-phase with each other, an important check against possible spurious effects.

(c) In the first part of this Problem, we estimated the density of dark-matter particles. One might ask, given this density and the fact that the particles should move with respect to a galaxy no faster than the escape velocity, is the gas of the dark-matter particles at *quantum degeneracy* or not? If the dark-matter particles are *fermions*, this would present a problem since there cannot be more than one such particle per quantum state. In other words, since the density is fixed by the observed dark-matter abundance, for fermions that are lighter than m^*, the *Fermi velocity* (equal to the *Fermi momentum* divided by m) exceeds the escape velocity, and the particles will leave the galaxy.

Let us assume, for example, that the dark-matter particles are spin-1/2 fermions. Accounting for two possible polarizations, the density of states is

$$\text{Density of quantum states} = \frac{2 \cdot 4\pi p^2 dp}{(2\pi\hbar)^3}, \tag{3.16}$$

where $p = mv$ is the momentum. The density of particles with momentum not exceeding the Fermi momentum p_F is

$$\frac{2 \cdot 4\pi p_F^3}{3 \cdot (2\pi\hbar)^3}, \tag{3.17}$$

which should equal that of Eq. (3.13) for $m = m^*$. We have:

$$\frac{8\pi (m^* 10^{-3} c)^3}{3 \cdot 8\pi^3 \hbar^3} = \frac{400 \text{ MeV/cm}^3}{m^* c^2}, \tag{3.18}$$

which yields

$$(m^* c^2)^4 = 400 \frac{\text{MeV}}{\text{cm}^3} 3\pi^2 10^9 (\hbar c)^3. \tag{3.19}$$

Evaluating the numerical values, gives $\boxed{m^* c^2 \approx 20 \text{ eV.}}$

Thus, if dark matter is made up of particles with rest-mass energy less than 20 eV, then these particles have to be bosons in order to remain gravitationally bound to galaxies.

DARK MATTER "THE ELEPHANT IN THE ROOM"

3.7   Detecting gravitational waves

The Laser Interferometer Gravitational-Wave Observatory (*LIGO*) experiment is a laser interferometer designed to detect *gravitational waves*. The interferometer consists of two arms, each $L = 4\,$km in length. When a gravitational wave passes through, one of the interferometer arms is stretched, and the other is compressed, which gives rise to a relative phase shift of the laser light traveling in the two arms. The wavelength of the laser light is $\lambda \approx 1\,\mu$m. In 2015, LIGO detected a gravitational wave from the merger of two black holes, whose masses are roughly 30 times the mass of the sun (Abbott et al., 2017).

(a) If a *black hole* has mass M, make a simple argument, based on non-relativistic Newtonian gravity, that its size (*Schwarzschild radius*) is given by

$$R_S = \frac{2GM}{c^2}. \tag{3.20}$$

What is the Schwarzschild radius of the Sun?[2]

(b) Consider a merger between two black holes, each with mass 30 times the mass of the Sun. Estimate their orbital velocity as they rotate around each other, at the moment when the separation between their centers is 10 times their size: $r = 10R_S$. Use the Newtonian approximation for gravitational force and ignore relativistic corrections.

(c) Estimate the frequency f of the orbital rotation of the two black holes around each other at this separation. Ignore relativistic corrections.

(d) The rotating black holes emit gravitational waves that are detected by the LIGO observatory on Earth at a distance $L = 500\,$Mpc away. The strength of the gravitational wave is characterized by dimensionless *strain* h, that quantifies the fractional change of distance between objects due to the gravitational wave. Einstein found that this strain at a distance L from the source is given by:

$$h = \frac{2G}{c^4 L} \frac{d^2 Q}{dt^2}, \tag{3.21}$$

where Q is the mass quadrupole moment of the source (Einstein, A., 1918; Abbott et al., 2017). We can approximate $d^2 Q/dt^2 \approx \omega^2 M r^2$, where $\omega = 2\pi f$.
    Estimate the numerical value of the strain h on Earth.[3]

(e) A gravitational wave incident on the LIGO interferometer causes a fractional change of the lengths of the two interferometer arms given by the strain $\Delta L/L = h$. The interferometer measures the light phase shift due to this change. During half of a gravitational wave oscillation, light travels up and down the interferometer arms 1000 times. What is the phase shift ϕ accumulated by the light due to the gravitational wave?

---

[2]In CGS units: $G \approx 7 \times 10^{-8}\,$cm^{3}g^{-1}s^{-2}, $c \approx 3 \times 10^{10}\,$cm/s, and mass of the Sun $M_\odot \approx 2 \times 10^{33}\,$g.
[3]In centimeter-gram-second (CGS) units: $1\,$pc $\approx 3.3\,$light $-$ years $\approx 3 \times 10^{18}\,$cm.

(**f**) Photon *shot noise* gives the fundamental limit on the uncertainty with which this phase can be measured. If the detector registers N photons, then the shot-noise limit on the phase uncertainty is[4]

$$\delta\phi = \frac{1}{\sqrt{N}}. \tag{3.22}$$

Estimate how many photons have to be detected in order to measure the phase shift ϕ due to the gravitational wave with signal-to-noise ratio of 30.

(**g**) What is the laser power (in watts) necessary to produce this number of photons?

---

[4]Advanced LIGO actually uses *squeezed light*, which improves the sensitivity beyond the photon shot-noise limit.

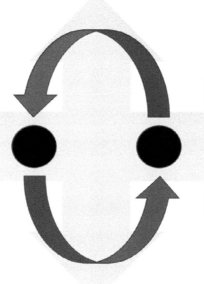

Solution

(a) In order to estimate the Schwarzschild radius of a black hole, we can find the distance from its center at which the escape velocity is equal to the speed of light:

$$v_e = \sqrt{\frac{2GM}{R_S}} = c,$$

which gives Eq. (3.20). This is a crude estimate that takes no account of relativistic effects, which are obviously important. As mentioned in Prob. 3.2, this amounts to dimensional analysis; the fact that it gives the correct numerical pre-factor is accidental.

The Schwarzschild radius of the Sun is: $\boxed{2GM_\odot/c^2 \approx 3 \,\text{km.}}$

(b) For black hole mass $M = 30M_\odot$, $R_S \approx 100 \,\text{km}$, thus $r \approx 1000 \,\text{km}$. Each black hole rotates around the center of mass of the system, so their orbital velocity can be estimated as

$$v = \sqrt{\frac{GM}{r/2}} = \sqrt{\frac{2GM/c^2}{r}}\, c,$$

which yields $\boxed{v \approx c/3.}$

(c) We can estimate the orbital period of the rotation around the center of mass as $2\pi r/2v$, and the frequency is the inverse:

$$\boxed{f = \frac{v}{\pi r} \approx 30 \,\text{Hz.}}$$

(d) Evaluating numerical values gives $\boxed{h = 10^{-21},}$ which is exactly the amplitude of the signal detected by LIGO (Abbott et al., 2017).

(e) The accumulated phase shift is

$$\phi = 1000 \times 2\pi \frac{\Delta L}{\lambda} = 1000 \times 2\pi \frac{hL}{\lambda} \approx \boxed{10^{-8}\,\text{rad.}}$$

(f) For a signal-to-noise ratio of 30 the shot-noise limit on the phase uncertainty has to be 3×10^{-10} rad, therefore the number of detected photons has to be $\boxed{N \approx 10^{19},}$ which, at wavelength of one micron, corresponds to 1 J of energy.

(g) The laser needs to deliver 1 J of energy every half period of the gravitational wave, which corresponds to laser power of approximately $\boxed{100\,\text{W.}}$

3.8 Dark matter trapped in the Earth

At the time of writing, little is known about the nature and composition of *dark matter*, and so a wide range of ideas and exotic scenarios are "on the table," until the day nongravitational interactions of dark matter are unambiguously observed. One property of dark matter that appears to be experimentally established is that it interacts with normal matter only weakly (if at all).

In one exotic scenario, a chunk of dark matter may occasionally find itself trapped in the gravitational potential of the Earth (we will not dwell here on how this might happen) and, having no nongravitational interactions with the planet's material, would potentially sink towards the center of the Earth.[5]

(**a**) What is the period of small radial oscillation of such a compact dark-matter object assuming it has no orbital angular momentum with respect to the center of the Earth? Make a simplifying assumption that the density of the planet is uniform (which is far from reality but gives a useful entry point to the problem).

(**b**) Assume now that a dark-matter chunk does have orbital angular momentum and, in fact, is on a circular orbit of radius $r < R_E$ (R_E is the radius of the Earth) around the center of the planet.

What is the period of revolution in this case?

(**c**) What is the period of motion for an arbitrary orbit fully confined within the uniform-density Earth?

[5]This problem was inspired by a discussion with Professor Allen P. Mills. We also acknowledge discussions with Professor Derek Jackson Kimball.

Solution

(a) The reader may have recognized a variation on a famous problem referring to an object thrown into a tunnel dug through the center of the Earth.

An object at a radius r from the center of the Earth only "feels" the mass enclosed in a sphere of the same radius. Assuming uniform density of the Earth, this enclosed mass can be written in terms of the Earth's mass M_E and radius R_E as

$$M = M_E \cdot \left(\frac{r}{R_E}\right)^3. \tag{3.23}$$

The gravitational acceleration towards the center of the Earth can be expressed through the gravitational acceleration on the surface g as

$$a = -g\frac{M}{M_E} \cdot \left(\frac{R_E}{r}\right)^2 = -g\frac{r}{R_E}. \tag{3.24}$$

We see that we have a *simple harmonic oscillator* here (a.k.a. *"mass on a spring"*), so we can immediately write the frequency as

$$\omega_0 = \sqrt{\frac{g}{R_E}}, \tag{3.25}$$

so that the oscillation period is

$$\boxed{T = \frac{2\pi}{\omega_0} \approx 80\,\text{min.}} \tag{3.26}$$

(b) The centripetal acceleration can be written as $g(r/R_E)$, so we have:

$$\frac{v^2}{r} = g\frac{r}{R_E}, \tag{3.27}$$

where v is the orbital velocity. From this, we have:

$$\boxed{T = \frac{2\pi r}{v} = \frac{2\pi}{\sqrt{g/R_E}},} \tag{3.28}$$

the same period as that in part (a).

(c) It probably will not come as a big surprise that the same answer would hold for *any* motion with a trajectory fully contained inside the Earth. This can be seen, for example, by writing the *Lagrangian* of the system:

$$L = T - V = \frac{p_x^2 + p_y^2 + p_z^2}{2m} + \frac{mg(x^2 + y^2 + z^2)}{R_E}, \tag{3.29}$$

where T is the kinetic energy, V is the potential energy (that we set to zero at infinity for convenience), p_i are the Cartesian components of momentum, and m is the mass of the dark-matter chunk. The motion with respect to each of the coordinates is governed by the *Lagrangian equations* that can be seen to be both independent and identical for all three coordinates, each of these motions being that of a harmonic oscillator. From this, it follows that the trajectories are closed, and the period of the motion is as derived in part (a).

4

Electromagnetism

Two hydrogen atoms walk into a bar. The first says to the second: "I think I lost my electron." The second asks: "Are you sure"? The first replies: "Yes, I'm positive."

Physics on Your Feet: Berkeley Graduate Exam Questions: or Ninety Minutes of Shame but a PhD for the Rest of Your Life! Dmitry Budker and Alexander O. Sushkov, Oxford University press. © Dmitry Budker, Alexander O. Sushkov, Vasiliki Demas 2015, 2021. DOI: 10.1093/oso/9780198842361.003.0004

4.1 Currents and magnetic fields

In this problem we will calculate magnetic fields created by current-carrying wires, and the forces between these wires (Fig. 4.1).

(**a**) What is the magnetic field outside an infinitely long straight wire carrying a current I?

(**b**) What is the magnetic field at the center of a circular loop of wire carrying a current I? The radius of the loop is R.

(**c**) Explain how one might be able to produce a current in a circular loop of wire, as in part (b).

(**d**) Evaluate the answer to part (b) if $I = 1$ A and $R = 1$ cm. Give the magnitude of the magnetic field in gauss.

(**e**) Calculate the force per unit length between two parallel wires, each carrying a current I, if they are separated by distance d. Evaluate the magnitude of this force for $I = 1$ A and $d = 100$ cm.

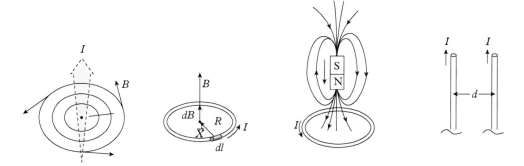

Fig. 4.1 Wires, currents, magnetic fields...

Solution

(a) We use the cylindrical coordinate system (r,ϕ,z), with the z-axis pointing along the wire in the direction of the current flow. Using *Ampere's law*:

$$\oint \mathbf{B} \cdot d\boldsymbol{\ell} = \frac{4\pi}{c} I, \qquad (4.1)$$

we solve for the magnetic field vector at distance r from the wire:

$$\boxed{\mathbf{B} = \frac{2I}{cr}\hat{\phi},} \qquad (4.2)$$

where, in this case, we denote $\hat{\phi} = \hat{\mathbf{z}} \times \hat{\mathbf{r}}$.

(b) According to the *Biot-Savart law*, the total magnetic field at any point P is given by the integral over the length of the loop

$$\mathbf{B} = \frac{1}{c} \int \frac{I d\boldsymbol{\ell} \times \hat{\mathbf{x}}}{x^2}, \qquad (4.3)$$

where the vector $d\boldsymbol{\ell}$ refers to an infinitesimal current-carrying segment of the wire and points in the direction of the current flow, the vector $\mathbf{x}$ points from this segment to the point P, and $\hat{\mathbf{x}}$ is the corresponding unit vector. Let us once again use cylindrical coordinates, so that the current is running in the direction of $\hat{\phi}$, and the origin is at the center of the loop. Since we want the magnetic field at the origin, $\hat{\mathbf{x}} = -\hat{\mathbf{r}}$, and the magnitude of $\mathbf{x}$ is R. Thus

$$\mathbf{B} = \frac{I}{c} \int \frac{\hat{\phi} \times (-\hat{\mathbf{r}})}{R^2} d\ell = \frac{I\hat{\mathbf{z}}}{cR^2} \int d\ell. \qquad (4.4)$$

The remaining integral is just the length of the loop circumference, so

$$\boxed{\mathbf{B} = \frac{2\pi I}{cR}\hat{\mathbf{z}}.} \qquad (4.5)$$

(c) The most obvious way to create a current in a piece if wire is to connect its two ends to a battery. But a wire loop has no ends! According to Maxwell's equations, a time-varying magnetic field creates an electric field, which produces a force on the charges, and therefore a current. To set up a current in a closed loop of wire, we have to apply a time-varying magnetic flux through this loop, which creates an *electro-motive force*, driving a current in the loop.

(d) This book uses CGS units, and the CGS unit of current is *statampere*. However almost nobody measures current in statamperes. To convert between amperes and statamperes, all we need to remember is that the *elementary charge* is 4.8×10^{-10} statcoulombs (CGS unit of charge) and 1.6×10^{-19} coulombs (SI unit of charge). Thus

1 ampere = 1 coulomb per second = 3×10^9 statamperes. Substituting $I = 1$ A (times the above conversion factor) and $R = 1$ cm into Eq. (4.5) gives

$$\boxed{B = 0.6 \text{ gauss.}} \tag{4.6}$$

(e) The force on a current-carrying segment $d\ell$ in a magnetic field **B** is given by

$$d\mathbf{F} = \frac{I d\ell \times \mathbf{B}}{c}. \tag{4.7}$$

Using the result of part (a), we write down the magnitude of the force per unit length acting on each of the two parallel wires:

$$\boxed{F/L = \frac{2I^2}{c^2 d}.} \tag{4.8}$$

This force is directed perpendicularly towards the other wire, and is attractive if the currents flow in the same directions, and repulsive if they flow in opposite directions.
    For $I = 1$ A $= 3 \times 10^9$ statamperes and $d = 100$ cm, the magnitude of this force is:

$$\boxed{F/L = 2 \times 10^{-4} \text{ dyne/cm} = 2 \times 10^{-7} \text{ newton/m.}} \tag{4.9}$$

This is precisely how an ampere is defined in the SI system of units: one ampere is that constant current which, if maintained in two straight parallel conductors of infinite length, of negligible circular cross section, and placed 1 meter apart in vacuum, would produce between these conductors a force equal to 2×10^{-7} newtons per meter of length.

4.2 Electromagnet design

A simple electromagnet is constructed by winding coils of insulated metal wire onto a bobbin as shown in Fig. 4.2. Let us assume that the radius of each turn of the wire is approximately the same, which is true when the thickness of the winding is much smaller than the radius of the coil. Suppose that we want to produce a certain constant magnetic induction of magnitude B in the center of the coil, and we wish to minimize the power dissipation due to resistive heating of the coil.

(a) Assume that you have available wire of a certain diameter (*gauge*). Argue that it makes sense to wind as many turns as will fit onto the bobbin.

(b) How does the dissipated power scale with the diameter of the wire?

(c) Given the result of part (b), what do you think are the considerations for the choice of the wire gauge? What else should the magnet designer be concerned about?

Fig. 4.2 A simple electromagnet is constructed by winding turns of insulated wire onto a bobbin.

Solution

(a) Suppose we can accomplish the design goal by winding the full coil (that has resistance R) and drawing a certain current I through it. If we wind fewer turns, let us say, a half, then we would need to draw twice the current to achieve the same induction. Then, the power dissipated in this case, $(2I)^2(R/2) = 2I^2R$, is twice the power dissipated in the case of the full coil. Thus, we see that, indeed, it makes sense to wind the full coil if we wish to minimize power dissipation.

(b) The number of turns in the full coil scales as d^{-2}, where d is the diameter of the wire. The resistance goes as the number of turns times the resistance of one turn, which also scales as d^{-2}, so that $R \propto d^{-4}$. The required current, as we determined in part (a), scales inversely with the number of turns, i.e., $I \propto d^2$, so that the dissipated power $I^2R \propto d^4 d^{-4}$ is independent of the diameter of the wire.

(c) Given the insight of part (b), the principal consideration for the choice of the wire diameter is the availability of an appropriate power supply. The power, as we have seen, is fixed, so the product of voltage and current is also fixed by fixing B and the coil geometry.

Among the factors that need to be considered in the design of the magnet (beyond choosing the wire gauge) are providing efficient cooling for the coil to dissipate the released heat, as well as taking into account possible increase of the resistance of the coil due to heating of the wire.

4.3 Field in a shield (with a coil)

Suppose we have a cylindrical shell made out of *high-permeability magnetic material*. For the purpose of the problem, we will assume an infinitely long cylinder and infinite permeability μ.

We wish to create a uniform transverse magnetic field within the cylinder as shown in Fig. 4.3. To achieve this, we are allowed to run current-carrying wires along the inner surface of the cylinder, in a thin layer adjacent to the wall, with currents flowing parallel to the cylinder's axis.

(a) Find the distribution of the current necessary to produce a certain value of the uniform magnetic field B.

(b) Now suppose that we keep the current distribution the same as in (a) but remove the magnetic material. How will the field change within the cylinder?

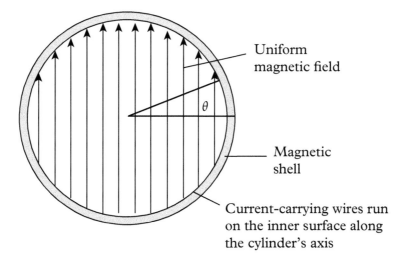

Fig. 4.3 We wish to create a uniform transverse magnetic field inside an infinitely long cylinder made of an infinite-permeability magnetic material.

Solution

(a) Inside the infinite-permeability material, we have $\mathbf{H} = 0$, where $\mathbf{H}$ is the magnetic field. On the other hand, the boundary conditions at the surface of the material tell us that the tangential component of $\mathbf{H}$ must be continuous. This can be satisfied by creating a surface current. From the Maxwell's equation

$$\nabla \times \mathbf{H} = \frac{4\pi}{c} \mathbf{j}, \tag{4.10}$$

where $\mathbf{j}$ is the current density, we get that a surface current with linear density of $j\delta$ (where δ is the thickness of the current layer) creates a step in the tangential component of $\mathbf{H}$ of $2\pi j\delta/c$.

The tangential component of the magnetic field inside the cylinder is $H_{\parallel} = B_{\parallel} = B\cos\theta$, which tells us that the sought for current distribution is

$$\boxed{j\delta = \frac{cB}{2\pi}\cos\theta.} \tag{4.11}$$

Due to the characteristic current distribution necessary to create a uniform field within a cylinder, the corresponding magnetic coils are called *cosine theta coils*.

(b) In order to answer this question, let us first solve the following problem: in the absence of the magnetic shield, what current distribution would we need at the surface of a cylinder in order to have a uniform transverse field within the cylinder?

From symmetry, we can guess that the field outside the cylinder will be that of a two-dimensional dipole $\boldsymbol{\mu}$:

$$\mathbf{B}_{out} = -\frac{\boldsymbol{\mu}}{r^2} + \frac{2(\boldsymbol{\mu} \cdot \mathbf{r})\mathbf{r}}{r^4}, \tag{4.12}$$

where $\mathbf{r}$ is the radius vector to the point where the field is evaluated in a plane transverse to the cylinder's axis.

Our "educated guess" can be verified by matching the boundary conditions at the surface of the cylinder. Once we do this, we find that, if the field inside the cylinder is of magnitude B_{in}, then, as before, the component of this field that is parallel to the surface is $B_{in\parallel} = B_{in}\cos\theta$, while the field outside has tangential component $B_{out\parallel} = -B_{in}\cos\theta$. Note that the second term in Eq. (4.12) does not contribute to the tangential component.

This argument shows that the current distribution that is needed to maintain this field configuration is actually exactly the same $\cos\theta$ distribution as in the case of a cylindrical shield. However, since the tangential field just outside the coil is equal and opposite to the tangential field just inside (as opposed to being zero in the case of the shield), if we keep the same current distribution, we will have half the field inside the coil compared to the case with the shield.

Another way to approach this problem is using the method of *image currents*. Let us go back to the case of the shield and imagine that there is a vanishingly thin uniform gap between the coil (assumed infinitely thin) and the shield. From the continuity of the tangential field, we see that the tangential field in this gap is zero. What this means is that there is a current flowing on the inner surface of the shield with exactly the same distribution and magnitude as the current in the coil. Thus, from the point of view of the field inside the shield, the effect of the shield is to double the coil current. Once the shield is removed, the effective current and the field inside the cylinder are both reduced by a factor of two.

4.4   Multipole expansion

Consider a localized system of static charges.

(a) Explain why the spatial distribution can always be characterized by a set of moments: *monopole* (total charge), *dipole*, *quadruple*, etc. Why are these 2^n-poles (where n is a non-negative integer)? (In other words, why do not we talk about *tripole*, *pentapole*, etc. moments?)

(b) Derive the formula for the electric field from a point dipole.

Solution

(a) If we have a point charge or, equivalently, are interested in the electrostatic potential far from the charges, then the electrostatic potential is

$$\varphi_0 = \frac{q}{r},\tag{4.13}$$

where the subscript indicates the lowest-order approximation, q is the total charge, r is the distance from the charge to the point of observation, and the potential is set to zero at infinite distance from the charge. The exact value of the potential at the observation point, taking into account the spatial distribution of the charges, can be written using the *Taylor expansion* of $1/r$, where moments of the distribution (which are, generally, tensors) are related to the integrals of the charge density with the corresponding powers of the coordinates $\mathbf{r}'$ (see Sec. 1-7 of Panofsky and Phillips 2005). For example, the Cartesian component of the dipole moment (which is a vector, i.e., a rank-one tensor) is

$$d_i = \int \rho(\mathbf{r}')r_i'dV',\tag{4.14}$$

where $\rho(\mathbf{r}')$ is the charge density, V' is a volume containing the charges, and for the *quadrupole moment*, we have:

$$Q_{ik} = \int \rho(\mathbf{r}') \left[3r_i'r_k' - r^2\delta_{ik}\right] dV'.\tag{4.15}$$

The terminology of 2^n-poles comes from the minimal way one can build a "pure" multipole from point charges. If we have one point charge, the 2^0 (monopole) moment is the only characteristic we have. Now, let us zero the monopole by taking two $(= 2^1)$ opposite charges q and $-q$. If they overlap, we have no electric field. If, however, we displace the charges by a small amount compared to the distance to the observation point, we have a dipole. The next thing we need to build is a quadrupole because it corresponds to the next term in the Taylor expansion [Eq. (4.15)]. It can be similarly built by taking two opposite dipoles (i.e., $2^2 = 4$ point charges) and displacing them a little with respect to each other. Note that, by construction, this procedure ensures that the distribution we build has no lower or higher moments. The latter is due to our judicious choice of small displacements.

Now, what about tripoles, pentapoles, and the such? Of course, nobody prevents us from constructing such groups of charges. However, whenever we do this, we will be able to *reduce* such distributions to combinations of "legitimate" multipoles. For example, three closely positioned charges generally correspond to a combination of a net charge and a dipole.

(b) Being proportional to both the *dipole moment* $\mathbf{d}$ and the first spatial derivative (i.e., *gradient*) of $1/r$, the scalar potential of the dipole field is

$$\varphi_1 = \frac{\mathbf{d} \cdot \mathbf{r}}{r^3}.\tag{4.16}$$

The electric field is then:

$$\boxed{\mathbf{E}_1 = -\boldsymbol{\nabla}\varphi_1 = -\frac{\mathbf{d}}{r^3} + \frac{3(\mathbf{d}\cdot\mathbf{r})\mathbf{r}}{r^5}.} \tag{4.17}$$

Note that the dipole-field formula is often needed in times of examinations (as well as on other occasions), and this is an excellent way to reconstruct it without much prior memorization.

Multipole personality

4.5   Energy in a wire

A long wire of radius R and conductivity σ carries a steady electrical current I.

(a) What are the electric ($\mathbf{E}$) and magnetic ($\mathbf{B}$) fields everywhere?

(b) Because of the finite conductivity, the wire heats up, and the heat dissipates into the surroundings. Where does the heating energy come from?

Solution

(a) Using cylindrical coordinates (r, ϕ, z) with $\hat{z}$ parallel to the current and centered on the wire, treating the wire as infinitely long and straight, cylindrical symmetry may be used with Ampere's law

$$\oint \mathbf{B} \cdot d\boldsymbol{\ell} = \frac{4\pi}{c} \int \mathbf{j} \cdot d\mathbf{A} \tag{4.18}$$

(ℓ and A are the circulation circuit length and area; $\mathbf{j}$ is the current density) to determine the magnetic field

$$\boxed{\mathbf{B} = \frac{2I\hat{\phi}}{cR^2} \times \begin{cases} r & : r \leq R \\ \frac{R^2}{r} & : r \geq R. \end{cases}} \tag{4.19}$$

The electric field inside the wire is

$$\boxed{\mathbf{E}_{in} = \mathbf{j}/\sigma = \frac{I}{\pi R^2 \sigma} \hat{z}.} \tag{4.20}$$

To find the field outside the wire, electric-field *circulation* around a loop that runs along the current in the wire and returns back outside may be considered, to get, taking account of symmetry,

$$\oint \mathbf{E} \cdot d\boldsymbol{\ell} = 0 \quad \Rightarrow \quad \boxed{\mathbf{E}_{out} = \mathbf{E}_{in} = \frac{I}{\pi R^2 \sigma} \hat{z},} \tag{4.21}$$

see also Eq. (A.11). So far, this is just a standard "E&M" (i.e., electricity and magnetism) problem.

(b) The simplest posed questions are sometimes the trickiest to answer. In this case, the tempting statement "the energy to heat the wire comes from the power supply via the wire," though ultimately correct (at least, in the authors' opinion), may be subject to rightful criticism in the following way.

Let us look at the energy balance in a given volume (assuming non-polarizable media)

$$\frac{d}{dt} \frac{1}{8\pi} \int \left(E^2 + B^2 \right) dV = \underbrace{\oint \mathbf{S} \cdot d\mathbf{A}}_{\text{rate of energy influx}} - \underbrace{\int \mathbf{E} \cdot \mathbf{j} \, dV.}_{\text{rate work is done on charges}} \tag{4.22}$$

$$\underbrace{\phantom{\frac{d}{dt} \frac{1}{8\pi} \int \left(E^2 + B^2 \right) dV}}_{\text{energy of the fields}}$$

Here V is the volume with surface-area elements $d\mathbf{A}$; $\mathbf{S}$ is the *Poynting vector*. Since the system is in steady state, the left-hand side of Eq. (4.22) is zero. The Poynting vector $\mathbf{S}$ is

$$\mathbf{S} = \frac{c}{4\pi} \mathbf{E} \times \mathbf{B} = -\frac{\hat{\mathbf{r}}}{2\sigma} \left(\frac{I}{\pi R^2} \right)^2 \times \begin{cases} r & : r \leq R \\ \frac{R^2}{r} & : r \geq R. \end{cases} \tag{4.23}$$

Because $\hat{\mathbf{r}}/r$ is divergence-free, any volume that lies entirely outside the wire trivially satisfies Eq. (4.22). Note that the direction of the Poynting vector (which is the direction of energy flow of the electromagnetic field) is towards the wire.

Within the wire, a cylindrical volume of radius r and length ℓ is considered to calculate

$$\oint \mathbf{S} \cdot d\mathbf{A} = S\big|_r \, 2\pi r \ell = \frac{1}{\sigma} \left(\frac{I}{\pi R^2} \right)^2 \pi r^2 \ell = \int \mathbf{E} \cdot \mathbf{j} \, dV. \qquad (4.24)$$

Thus the work of Joule heating is balanced by the flux of electromagnetic energy into the volume. It appears that the energy is coming from the fields, and the direction of energy transport is given by the Poynting vector, i.e., it enters the wire radially from the outside! This is a "macroscopic" picture.

So is it wrong to say that the energy comes from the power supply and is carried to the place where it is converted to heat by electrons? Certainly, the fields are established as a result of the action of the power supply. As for electrons, they constantly lose their *kinetic energy* gained from acceleration between collisions within the wire, so the kinetic energy of the electrons does not change on average as they move with the current. However, what the electrons do lose moving down the current is their *potential energy*. A quick calculation shows that the amount of potential energy lost is exactly the same as the amount of generated heat.

This example illustrates a not-too-uncommon situation in physics: there sometimes exist rather different descriptions of the same effect, seemingly unrelated to each other. (A famous example is the *Casimir effect* that consists in existence of a force between grounded conducting objects. In one picture, the force arises due to the interaction of fluctuating dipoles in the conductors, and in the other—it is due to the *vacuum fluctuations* of the electromagnetic field between the conductors.) Then, physicists may face a dilemma: are the different descriptions redundant, or do the different mechanisms act independently?

In this case, the problem is simple enough, and the answer is unambiguous: the two descriptions are complementary, and only one picture should be used at a time to avoid *overcounting*.

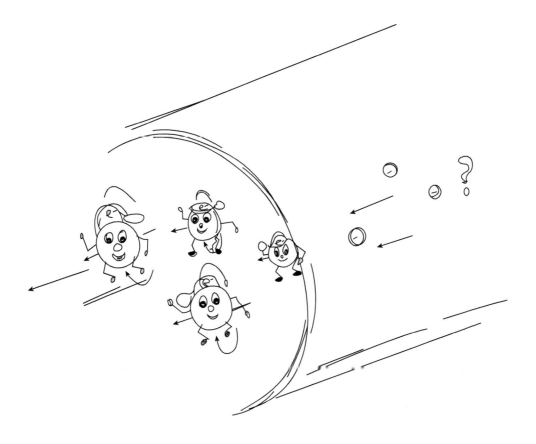

4.6 Earth's magnetic field angle

Neglecting the fact that the magnetic poles do not exactly coincide with the geographic poles and making other reasonable simplifying assumptions about the Earth's magnetic field, determine the angle the Earth's magnetic field makes with the normal to the Earth's surface at a location with a given latitude.

> In fact, the approximations we will be making in this problem are quite crude. For example, the horizontal component of the Earth's magnetic field at Berkeley is tilted by about 15° with respect to the "nominal" North-to-South direction (this is called *magnetic declination*). To make matters worse, magnetic declination changes in time, in an irregular way, with a variation of degrees per decade in certain locations, see, for example, `http://geomag.nrcan.gc.ca/mag_fld/magdec-en.php`.

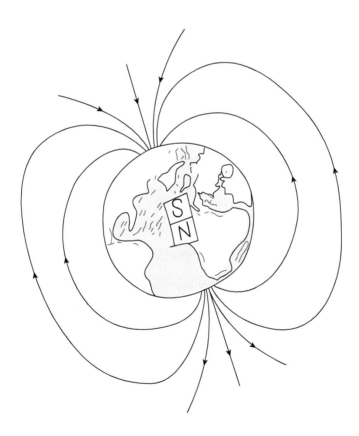

Solution

We will approximate the Earth's magnetic field as that of a dipole $\boldsymbol{\mu}$ located at the Earth's center and oriented along the line connecting the geographical poles. We will derive the expression for the dipole magnetic field at a distance r from the center starting from an easy-to-remember expression for the *scalar magnetic potential*

$$\varphi = \frac{\boldsymbol{\mu} \cdot \mathbf{r}}{r^3}. \tag{4.25}$$

The magnetic field is found according to:

$$\mathbf{B} = -\boldsymbol{\nabla}\varphi = -\frac{\partial \varphi}{\partial \mathbf{r}} = -\frac{\boldsymbol{\mu}}{r^3} + \frac{3(\boldsymbol{\mu} \cdot \mathbf{r})\mathbf{r}}{r^5}, \tag{4.26}$$

where we took into account that

$$\frac{\partial r}{\partial \mathbf{r}} = \frac{\partial \left(x^2 + y^2 + z^2\right)^{1/2}}{\partial \mathbf{r}} = \frac{\mathbf{r}}{r}. \tag{4.27}$$

The second term in Eq. (4.26) is along the vertical and has a magnitude of

$$3\frac{\mu}{r^3} \cos\theta, \tag{4.28}$$

where θ is the angle between $\boldsymbol{\mu}$ and $\mathbf{r}$. The first term in Eq. (4.26) generally has both a vertical and a horizontal component. The vertical component is

$$-\frac{\boldsymbol{\mu} \cdot \mathbf{r}}{r^4} = -\frac{\mu}{r^3} \cos\theta. \tag{4.29}$$

The horizontal component is, correspondingly,

$$-\frac{\mu}{r^3} \sin\theta. \tag{4.30}$$

The angle α the total field comprises with the vertical is found from

$$\tan\alpha = \frac{-\sin\theta}{2\cos\theta} = -\frac{\tan\theta}{2}. \tag{4.31}$$

Finally, the problem is solved by identifying θ as 90° minus the latitude.

For our idealized situation, we find, as expected, that $\alpha = 0$ (vertical magnetic field) at the North Pole, and horizontal magnetic field at the equator. At the latitude of Berkeley ($\approx 38°$), we calculate that the magnetic field is inclined from the vertical by about 33°.

4.7   Refrigerator-magnet science

(a) What is the magnetic field **H** and magnetic induction **B** everywhere associated with a thin sheet of a *ferromagnetic material* uniformly magnetized perpendicular to the surface of the sheet?

(b) Same for a sheet magnetized parallel to its surface.

(c) Based on the results of parts (a) and (b), explain why *refrigerator magnets* are typically made out of *periodically poled* magnetic sheet, where magnetization is constant along one direction in the plane of the sheet but reverses with a spatial period commensurate with the thickness of the sheet in the other transverse direction.[1]

---

[1]This was a favorite exam problem of Prof. Max Zolotorev.

Solution

(a) Magnetization ($\mathbf{M}$) of a volume element can be thought of as a loop of *surface current* proportional to the magnitude of the magnetization and flowing around the element in a plane perpendicular to $\mathbf{M}$. If magnetization is uniform, then the currents from adjacent elements in the material cancel each other, except at the boundary of the material. Thus, the fields from a thin sheet is that of a current loop at the edge of the sheet. Since the field from a current loop drops with the dimension of the loop, we are led to conclude that, neglecting edge effects,

$$\boxed{\mathbf{B} = \mathbf{H} = 0 \text{ outside.}} \tag{4.32}$$

We should remark that many students find this result disturbing...

What about inside the sheet? First of all, we use the *boundary condition* of continuity of the normal component of $\mathbf{B}$ to conclude that the magnetic induction inside the material is also zero. Finally, using the general relation

$$\mathbf{B} = \mathbf{H} + 4\pi\mathbf{M}, \tag{4.33}$$

we find

$$\boxed{\mathbf{B} = 0; \quad \mathbf{H} = -4\pi\mathbf{M} \text{ inside.}} \tag{4.34}$$

(b) In this case, the magnetization is equivalent to opposite surface currents flowing at the top and bottom of the sheet. Since the magnetic field produced by these currents cancels outside of the sheet, we, once again, find:

$$\boxed{\mathbf{B} = \mathbf{H} = 0 \text{ outside.}} \tag{4.35}$$

From the continuity of the tangential component of $\mathbf{H}$, we now also find that $\mathbf{H} = 0$ inside the material, and, finally,

$$\boxed{\mathbf{B} = 4\pi\mathbf{M}; \quad \mathbf{H} = 0 \text{ inside.}} \tag{4.36}$$

(c) From the results of parts (a) and (b), we see that uniformly magnetized sheets will not do well as refrigerator magnets as they do not produce any magnetic field outside of them apart from those near the edge. From our first-hand experience, if you take a sheet of uniformly transversely magnetized flexible magnetic material with 1 mm thickness and cut a piece with dimensions of a typical refrigerator magnet (8 cm ×6 cm, for instance), such a magnet would barely stick to a refrigerator wall.

In a periodically poled magnet, field lines are not confined within the sheet, and such magnet is nicely attracted to an iron wall of a refrigerator.

A pair of refrigerator magnets is a wonderful educational toy nicely illustrating what we said above (and more). If the two magnets are put together so they are touching on the sides that are supposed to be facing a refrigerator, they stick to one another, and one finds that sliding the two magnets with respect to each other is smooth if the motion is along one edge, while it is "jerky" if the magnets are displaced in the orthogonal direction. This is a demonstration of periodic poling.

It turns out, turning the magnets around so they face each other on the sides that are supposed to be away from the refrigerator yields another surprise: the magnets barely "feel each other" at all! The reason for this is a special pattern of periodic poling known as the *Halbach array* that essentially doubles the field on one side and cancels it on the other. Can you think of a magnetization pattern that would do this?

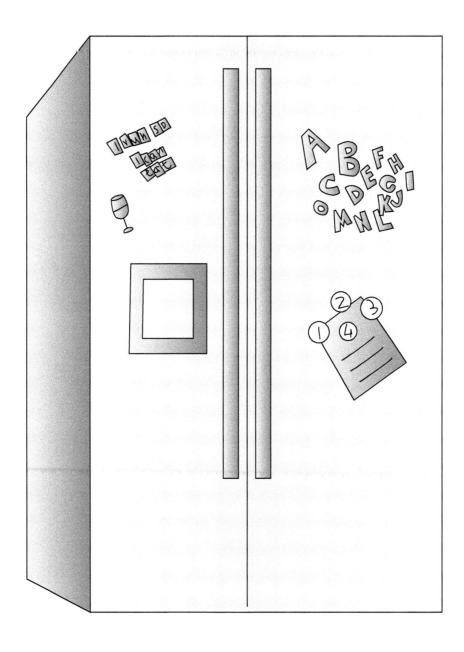

4.8   Spherical-cell magnetometer

The sensitive volume of a *magnetometer* is a ball of radius R (Fig. 4.4). The magnetometer reports each of the three Cartesian components of the external magnetic induction, B_x, B_y, and B_z, averaged over the sensitive volume. A point dipole $\boldsymbol{\mu} = \mu \cdot \hat{\mathbf{z}}$ is placed at a distance $2R$ along $\hat{\mathbf{x}}$ from the center of the magnetometer.

What does the magnetometer read?

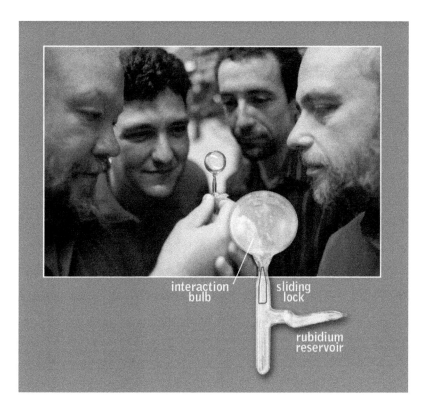

Fig. 4.4 Physicists examining a vapor cell for a rubidium-vapor magnetometer. Left-to-right: Drs. M. Balabas, T. Karaulanov, M. Ledbetter, and D. Budker. Berkeley, 2010. Photo courtesy of Dr. Damon English.

Solution

The magnetic induction produced by the dipole at the center of the sensor volume is $B_x = B_y = 0$ (from symmetry), and $B_z = -\mu/(2R)^3$ (from the dipole-field formula).

A remarkable property of the static magnetic field, also shared by static electric and gravitational fields, is that, if the sources of the field are outside a sphere, the field averaged either over the surface of the sphere or over its volume is equal to the value of the field at the center. Therefore, the result of the previous paragraph is, in fact, also the problem's answer.

The result for the case of the gravitational field is usually introduced in the very first physics course. For example, to calculate the total gravitational force from the Moon on the Earth, we can assume that both are point masses concentrated in their respective sphere centers. The property can be proven by explicit integration, but it also follows directly from the *average-value theorems* for *Harmonic functions*, i.e., the solutions of the *Laplace equation*. If one adopts the direct-integration approach, it is sufficient to first consider a single point charge (or mass, or *magnetic monopole*) outside the sphere, and verify the result for this simple case. Then, the general result follows from the *superposition principle*, because the field (induction) inside the sphere can be thought of as originating from a set of charges outside of it.

4.9   Magnetic force on a superconducting magnet

In this problem we will consider forces due to static electric and magnetic fields.

(a) Consider an infinite parallel-plate capacitor with electric field E between the plates. What is the force per unit area (pressure) between the plates?

(b) Consider a pair of infinite conducting sheets that carry currents in opposite directions, so that the magnetic field between the sheets is B. What is the force per unit area (pressure) between the plates?

(c) Calculate the pressure due to the magnetic force on the windings of a superconducting magnet with magnetic induction B in the bore.

Yield strength of structural steel is about 200 MPa. What is the maximum magnetic induction that can be produced in a *superconducting magnet* before the steel supports buckle?

Solution

(a) Consider a single thin infinite conducting plate with charge density per unit area σ. Applying Gauss's law gives the electric field on both sides of the plate: $E_1 = 2\pi\sigma$. Now suppose there is a second, oppositely charged, plate at some distance from the first. The electric fields due to the two plates add in such a way that the field on the outside of the capacitor is zero, and the field between the plates is $E = 4\pi\sigma$.

Now let us consider the force between the plates. The first plate applies on the second plate a force per unit area $F/A = \sigma E_1 = 2\pi\sigma^2$. Note that here we take the field E_1 due to the first plate, and not the total electric field inside the capacitor, since the second plate can not exert a force on itself. Expressed in terms of the total electric field E, this pressure is given by

$$P = F/A = -\frac{E^2}{8\pi}. \tag{4.37}$$

The absolute value of this pressure is equal to the energy density due to electric field E, as can be seen by considering the work that has to be done to pull the plates apart.

(b) Consider a single thin infinite conducting plate with a current density (current per unit length) κ. Applying Ampere's law gives the magnetic field on both sides of the plate: $B_1 = 2\pi\kappa/c$. Now suppose there is a second plate at some distance from the first, with the same current running in the opposite direction. The magnetic fields due to the two currents add in such a way that the field on the outside of the plates is zero, and the field between the plates is $B = 4\pi\kappa/c$.

Now let us calculate the force between the plates. Consider a square section of the second plate, with linear size x. The force on this section due to the magnetic field B_1 created by the first plate has a magnitude of $F = \kappa x^2 B_1/c$ and it is directed in such a way that it repels the plates. Note that here, once more, we take the field B_1 due to the first plate, and not the total magnetic field between the plates, since the second plate can not exert a force on itself. Expressed in terms of the total magnetic field B, the force per unit area, or pressure, is

$$P = F/A = \frac{B^2}{8\pi}. \tag{4.38}$$

The expression on the right-hand-side is the energy density due to magnetic field B. Note the sign is different from that in Eq. (4.37), this is related to the work that has to be done by the current supply to run the constant currents in the sheets, if they were to be pulled apart, for example.

(c) Equation (4.38) gives the pressure exerted on the windings of a superconducting magnet. The pressure is positive, so it wants to make the magnet explode. It is useful to remember that a 1 T magnetic field exerts a magnetic pressure of approximately 4 atm.

The atmospheric pressure is 10^5 Pa, therefore 200 MPa corresponds to 2000 atm. Such a pressure is produced by a magnetic field with induction of about 22 T. Most modern *NMR superconducting magnets* (NMR stands for nuclear magnetic resonance) run at magnetic inductions between 10 and 21 T. The magnetic field of the largest magnets is indeed limited by the mechanical pressure exerted by the magnetic field on the windings. Note that magnets with stronger DC fields (currently up to ~45 T for DC magnets) can be designed that employ redistribution of the mechanical pressure over the thickness of the coil. Another important design factor is the critical field of the superconducting windings, which is on the same order, and depends on the winding material.

Readers interested in the current state of the art of DC and pulsed high-field magnets for fundamental-physics research are referred to a review article by Battesti et al. (2018).

4.10 Circuit view of atoms and space

Thinking of physical systems as electric circuits can lead to useful insights. In this problem, we discuss free space as a *transmission line*, and an atom as an *LC circuit* and an *antenna*.

(**a**) *Wave impedance* of an electromagnetic wave in a medium or a *waveguide* is defined as the ratio of the electric- and magnetic-field amplitudes in the wave, perhaps, up to a factor of $4\pi/c$, depending on the system of units used [see Appendix on units in the book by Jackson (1998)].

Show that the *impedance of free space* is $Z_0 \approx 377$ ohm when expressed in SI units. Hint: the Gaussian unit of resistance is $1\ \text{s/cm} = 9 \cdot 10^{11}$ ohm.

(**b**) Thinking of an atom as a *lumped LC circuit*, find the characteristic oscillation frequency. Compare this frequency with a characteristic frequency of electronic excitations in an atom.

(**c**) Consider an atom as an *antenna* emitting *electric-dipole radiation*. What is the *radiation resistance* of such "atomic antenna", i.e., the proportionality coefficient between the radiated power and the square of the electric current in the antenna.

(**d**) Same for *magnetic-dipole radiation*.

Solution

(a) For free space, the ratio of electric- and magnetic-field amplitudes in the wave expressed in Gaussian units is unity. Thus, we have:

$$Z_0 = 1 \;\rightarrow\; 4\pi/c \;\rightarrow\; \approx \frac{9 \cdot 10^{11} \ \text{ohm} \cdot \text{cm/s}}{(3 \cdot 10^{10})/(4\pi) \ \text{cm/s}} \approx 377 \ \text{ohm}. \tag{4.30}$$

(b) The capacitance C of a sphere is equal, in Gaussian units, to its radius. In this case, we can set the radius equal to the Bohr radius, co that $C = a_0$. To find the characteristic self-inductance L, we use the fact that

$$\Phi = cLI, \tag{4.40}$$

where $\Phi \sim \mathbf{B} \cdot \mathbf{A}$ is the flux of magnetic induction $\mathbf{B}$ through a contour of area $\mathbf{A}$, and I is the current. Here the speed of light c appears in the numerator because we use Gaussian units where inductance is measured in s^2/cm. Substituting, for an estimate,

$$\mathbf{B} \sim \frac{2\pi I}{a_0 c}, \tag{4.41}$$

as for the magnetic induction at the center of a circular loop, and $A \sim \pi a_0^2$, we get

$$L \sim \frac{2\pi^2 a_0}{c^2}, \tag{4.42}$$

so that, ignoring the numerical factor, we get for the characteristic frequency of the LC circuit:

$$\omega^* = \frac{1}{\sqrt{LC}} \sim \frac{c}{a_0}. \tag{4.43}$$

The characteristic frequency of electronic excitation is

$$\omega_0 = \frac{Ry}{\hbar} \sim \frac{e^2}{\hbar a_0} = \alpha \frac{c}{a_0} = \alpha \omega^*. \tag{4.44}$$

Here Ry is the Rydberg constant, and $\alpha = e^2/(\hbar c)$ is the fine-structure constant. Therefore, the characteristic frequency of electronic excitations in an atom is $\alpha \approx 1/137$ times that of the "atomic LC circuit".

(c) The power radiated by an electric dipole is (see, for example, Heald and Marion 1995):

$$P = \frac{2}{3c^3}\ddot{d}^2, \tag{4.45}$$

where d is the dipole moment. In the spirit of the estimate, we set $d \sim e a_0$, where e is the charge. If the dipole oscillates at frequency ω, then the current is $I \sim \omega q$, which, substituting into Eq. (4.45) and ignoring the numerical coefficient, yields

$$P \sim \frac{\omega^2 a_0^2}{c^3} I^2, \tag{4.46}$$

so that for the radiation resistance we get

$$R(E1) \sim \frac{\omega^2 a_0^2}{c^3} = \frac{4\pi^2}{c} \left(\frac{a_0}{\lambda}\right)^2 \sim \left(\frac{a_0}{\lambda}\right)^2 Z_0, \tag{4.47}$$

where λ is the radiation wavelength assumed much larger than a_0 in accordance with our finding above that $\omega_0 \sim \alpha \omega^*$, which means that $a_0 \sim \alpha \lambda$. Thus, for typical atomic transitions, the radiation resistance of an atom is $\sim \alpha^2 Z_0$, so the atomic antenna is poorly impedance matched to free space.

(d) In the case of magnetic-dipole radiation, we model an atom as a current loop of radius a_0, and calculate the radiated power using Eq. (4.45) substituting the electric dipole moment with the magnetic dipole moment of the current loop

$$\mu = \frac{IA}{c} = \frac{I\pi a_0^2}{c}. \tag{4.48}$$

This yields:

$$R(M1) \sim \frac{\omega^4 a_0^4}{c^5} \sim \left(\frac{a_0}{\lambda}\right)^4 Z_0, \tag{4.49}$$

i.e., even smaller radiation resistance (for a given radiation wavelength).

4.11   Magnetic monopole

A *magnetic monopole* is a hypothetical particle that possesses a magnetic rather than electric charge. It is appealing to consider the existence of magnetic monopoles because it would complete the symmetry between electric and magnetic fields that is present in the *Maxwell's equations* up to the charge and current terms.

Imagine a magnetic monopole (Fig. 4.5) slowly flying through a superconducting loop. Assuming that initially there is no current in the superconducting loop, describe the behavior of the current as a function of time ($t = 0$ corresponds to the time the monopole is in the plane of the loop).[2]

Fig. 4.5 A magnetic monopole (depicted as a *monopolist*) flying through a superconducting loop.

---

[2]This problem comes to us from Prof. J. D. Jackson.

Solution

Applying the *Gauss law* to the monopole, we find that the total *magnetic flux* through any closed surface surrounding the monopole is $\Phi_m = 4\pi g$, where g is the *magnetic charge* of the monopole. As the monopole approaches the loop, the flux of the monopole's field through the loop will gradually increase. When the monopole has nearly reached the plane of the loop, half of its field's flux is going through the loop.

Now, a superconducting loop cannot "tolerate" a change of total flux through it because this would cause a nonzero curl of the electric field and, consequently, a nonzero field within the superconductor. This is avoided by a current that flows around the superconducting loop and exactly compensates the external flux. The current is found from the relation

$$\Phi_s = \frac{LI}{c}, \tag{4.50}$$

where Φ_s is the magnetic flux through the loop due to its electrical current and L is the loop's self-induction. Thus the magnitude of the current just before the monopole has reached the loop is $I = 2\pi gc/L$.

Now, one might think that something weird happens when the monopole crosses the plane of the superconducting loop. Indeed, just after that moment, we, once again, have half of the total monopole's flux going through the loop, except the sign has changed! Does this mean that the sign of the current in the loop abruptly reverses?

If your intuition tells you that this cannot be, trust your intuition! Indeed, we have to remember how the Maxwell's equations look in the presence of monopoles. Specifically, let us write the curl equation for the electric field in the integral form. In the presence of the monopole, it is:

$$-\oint \mathbf{E} \cdot d\ell = \frac{d\Phi}{dt} + \frac{4\pi}{c} I_m. \tag{4.51}$$

Here I_m is the current of magnetic monopoles, and both sides should be identically zero for a superconducting loop.

In our case, the current of the monopoles piercing the superconducting loop is $I_m = g\delta(t)$. Let us write

$$\frac{d\Phi}{dt} = -\frac{4\pi}{c} I_m \tag{4.52}$$

and integrate both sides over time in the vicinity of $t = 0$. We see that when the monopole passes the plane of the loop, the apparent change in the flux discussed above is exactly compensated by the current term.

The bottom line is that when the monopole passes through the loop, there is smooth and monotonic increase of the current in the superconducting loop, leaving the loop with a finite current after the monopole is gone. This property has been used in experiments searching for magnetic monopoles (Huber et al., 1990).

We should also mention that there exist alternative approaches to the solution that lead to the same result. One approach uses the fact that the shape of the surface of the contour should not affect the result, so a convenient trick is to mentally deform the plane of the superconducting loop into a "bag" in such a way that the monopole would never cross the surface, thus avoiding the consideration of the (pseudo) discontinuity we discussed above. Another approach is to employ a model of a monopole where it corresponds to one end of a *semi-infinite solenoid* (Fig. 4.6). The narrow solenoid tube brings the magnetic flux that proceeds to disperse radially appearing as a magnetic charge. In this model, we can arrange things in such a way that once the end (i.e., the monopole) passes through the loop, the tail and its associated flux continue to pierce the plane of the loop, again yielding the same result for the superconducting current.

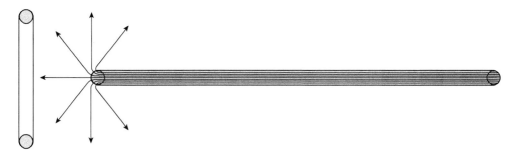

Fig. 4.6 A magnetic monoplole can be modeled as one end of a "semi-infinite solenoid."

5
Optics

_____

There are two types of people in the
world: those who can extrapolate
from incomplete information.

Physics on Your Feet: Berkeley Graduate Exam Questions: or Ninety Minutes of Shame but a PhD for the Rest of Your Life! Dmitry Budker and Alexander O. Sushkov, Oxford University press. © Dmitry Budker, Alexander O. Sushkov, Vasiliki Demas 2015, 2021. DOI: 10.1093/oso/9780198842361.003.0005

5.1   Rotating liquid mirror

What is the shape of the surface of liquid (e.g., mercury) in a container rotating around a vertical axis (Fig. 5.1) at a steady angular velocity ω? What is the focal length of a mirror formed by the liquid's surface? Comment on whether the deviation of the surface shape from spherical is detrimental or advantageous from the point of view of optical properties of the mirror.

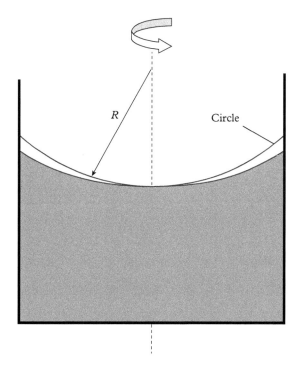

Fig. 5.1 The steady-state liquid surface in a container uniformly rotating around a vertical axis.

Solution

In the steady state, the liquid and the container rotate together as if they were one solid body. The centrifugal forces are pushing the liquid radially away from the rotation axis, and this causes the level of the liquid to rise towards the periphery of the container compared to that at the axis (Fig. 5.1). The problem is most conveniently analyzed in the *rotating frame*. In this frame each part of the liquid experiences a centrifugal force, which can be described by an effective *"centrifugal potential"* $-(wr)^2/2$, which can be seen by integrating the *centripetal acceleration* w^2r along a radial direction. The surface of the liquid in equilibrium must be an *equipotential*, which leads to the following equation for the height differential h as a function of the distance from the axis r:

$$gh = \frac{(wr)^2}{2}, \tag{5.1}$$

which shows that the surface of the liquid is a *paraboloid*.

We next find the radius of curvature R of the surface, making use of the fact that, near the axis, the paraboloid coincides with its tangential sphere. For the sphere, we have:

$$h = R - \sqrt{R^2 - r^2} \approx \frac{r^2}{2R}. \tag{5.2}$$

Comparing this expression with Eq. (5.1), we find that

$$R = \frac{g}{w^2}. \tag{5.3}$$

Finally, remembering that the focus of a curved mirror lies at $R/2$, we find the sought for expression for the focal distance:

$$\boxed{f = \frac{g}{2w^2}.} \tag{5.4}$$

The fact that the surface is a paraboloid rather than a sphere is beneficial because such a surface is free from *spherical aberrations*, meaning that a paraboloid will reflect a ray that is parallel to the axis through the focus, independently of how far from the axis the ray hits the mirror. For a sphere, this is only true for rays near the axis, i.e., in the *paraxial approximation*.

The method described in this problem is actually used to make telescope mirrors (with the focal length adjusted by changing the rotation rate), as well as for making large (eventually stationary) mirrors by gradually cooling molten glass in a rotating container until it solidifies.

5.2 Stacking lenses

In this problem, we discuss how to describe, in the framework of *geometrical optics*, single thin lenses, as well as periodic sequences of such lenses separated by free-space intervals, a contraption sometimes referred to as a *lensguide*. The material in this problem is discussed in much more detail in a number of excellent texts, of which our favorites are the books by Siegman (1986) and Nagourney (2014).

(a) In geometrical optics, we describe light propagation using *rays*. When a ray hits a thin lens a distance r_1 from the axis, it emerges on the other side with $r_2 = r_1$ right after the lens, however, the angle of the ray with respect to the axis is different: $\theta_2 \neq \theta_1$ (Fig. 5.2, left).

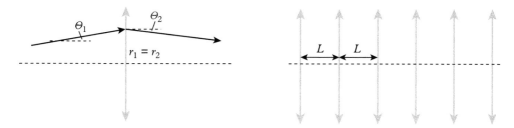

Fig. 5.2 Left: a ray is bent by a thin lens, but its distance from the axis remains unchanged at the lens. Right: a lensguide consisting of a sequence of positive and negative lenses.

Instead of dealing with the angles θ themselves, it is more convenient to use their tangents, which then directly relate to how the ray's distance from the axis changes with longitudinal coordinate z in free propagation:

$$r' = \frac{dr}{dz} = \tan\theta. \tag{5.5}$$

The two parameters r and r' completely define a ray at any given longitudinal position z. If a ray encounters an optical element, then we can relate the parameters of the ray just before and just after the element according to

$$\begin{pmatrix} r_2 \\ r_2' \end{pmatrix} = \begin{pmatrix} A & B \\ C & D \end{pmatrix} \begin{pmatrix} r_1 \\ r_1' \end{pmatrix}, \tag{5.6}$$

where the matrix for an element is referred to as the *ABCD matrix*.

Find the *ABCD* matrix for a thin lens.

(b) Find the *ABCD* matrix for a free-space interval of length L.

(c) What is the *ABCD* matrix that corresponds to a sequence of N elements, each consisting of a positive lens with focal length f, a free space of length L, a negative lens of focal length $-f$, and another free-space interval of length L?

(d) We will now address the question of *stability* of an infinite periodic optical system ($N \to \infty$). In other words, suppose we send an arbitrary paraxial ray onto the system.

Will it remain confined in the transverse direction as it propagates in the system, or will it "catastrophically" diverge from the axis?

In order to address this question, we will find the *eigenrays* r_a and r_b and the corresponding eigenvalues Λ of the $ABCD$ matrix of one period of the system. These eigenrays have the property that if we send such a ray at the input of a period, the ray at the output will be the same up to a constant multiplier, which is the corresponding eigenvalue Λ. An arbitrary input ray can be written as a sum of the two eigenrays with some coefficients. This is analogous to how we usually treat temporal evolution of a *two-level system* in quantum mechanics. Note that the eigenrays and eigenvalues are generally complex, although any physical ray should be real.

Find the eigenvalues of the $ABCD$ matrix of one period of the system, and based on the result, discuss its overall *stability criterion*.

Solution

(a) A thin lens in the *paraxial approximation*, where the angles of all rays are small with respect to the propagation direction, has the property that the incoming ray is bent by an angle proportional to r_1, so that, if the incoming ray is, for instance, parallel to z, then the bent ray goes through the *focus* at a *focal distance* f from the lens. We can write:

$$r_2 = r_1, \tag{5.7}$$
$$r_2' = -(1/f)r_1 + r_1', \tag{5.8}$$

or in the matrix form:

$$\begin{pmatrix} r_2 \\ r_2' \end{pmatrix} = \begin{pmatrix} 1 & 0 \\ -1/f & 1 \end{pmatrix} \begin{pmatrix} r_1 \\ r_1' \end{pmatrix}. \tag{5.9}$$

The 2×2 matrix in this expression is the sought after $ABCD$ matrix for a thin lens. Note that for a diverging lens, f is negative, and the formula still holds just fine.

(b) When a ray propagates through a free-space interval, its angle does not change, but the distance r changes according to:

$$r_2 = r_1 + r_1' L. \tag{5.10}$$

The corresponding $ABCD$ matrix is

$$\begin{pmatrix} 1 & L \\ 0 & 1 \end{pmatrix}. \tag{5.11}$$

(c) It follows from the definition of the $ABCD$ matrices that, in order to find the matrix corresponding to a composite element, we need to multiply the matrices for individual elements in reverse order. In this case, the matrix for one composite element which is a period of the system is:

$$\begin{pmatrix} 1 & L \\ 0 & 1 \end{pmatrix} \begin{pmatrix} 1 & 0 \\ 1/f & 1 \end{pmatrix} \begin{pmatrix} 1 & L \\ 0 & 1 \end{pmatrix} \begin{pmatrix} 1 & 0 \\ -1/f & 1 \end{pmatrix} = \begin{pmatrix} \frac{f+L}{f} & \frac{L(2f+L)}{f} \\ -\frac{L}{f^2} & \frac{f^2-fL-L^2}{f^2} \end{pmatrix}. \tag{5.12}$$

A sequence of N such elements is represented by the Nth power of this matrix.
    Evaluating the eigenvalues of the matrix (5.12), we find:

$$\Lambda_{a,b} = 1 - \frac{L^2}{2f^2} \pm \sqrt{\frac{L^4}{4f^4} - \frac{L^2}{f^2}} = m \pm \sqrt{m^2 - 1} \tag{5.13}$$

with $m = 1 - L^2/(2f^2)$. An interesting property of the eigenvalues seen by examining Eq. (5.13) is that $\Lambda_a \times \Lambda_b = 1$. This means that these two generally complex numbers have reciprocal amplitudes and conjugate phases.
    In order to analyze the stability of the periodic system, let us notice that, if we put in a ray $c_a r_a + c_b r_b$, we will get a ray represented by $c_a \Lambda_a r_a + c_b \Lambda_b r_b$ after the

first period, and $c_a \Lambda_a^N r_a + c_b \Lambda_b^N r_b$ after the Nth period. If the amplitude of one of the eigenvalues exceeds unity, the coefficient of the corresponding eigenray will increase exponentially with N, and the system is clearly unstable. On the other hand, if the amplitude of an eigenstate is unity, the magnitude of the coefficients for both eigenrays remains equal to unity with increasing N, meaning that the rays never diverge too far from the axis. Further examination of Eq. (5.13) shows that, indeed, if $|m| \le 1$ (which corresponds to $L \le 2f$), then Eq. (5.13) can be rewritten as

$$\Lambda_{a,b} = m \pm \sqrt{m^2 - 1} = m \pm i\sqrt{1 - m^2 -} = \cos\beta \pm i\sin\beta = e^{\pm i\beta}, \qquad (5.14)$$

where $m = \cos\beta$, β is real, and so the amplitudes of the eigenvalues $\Lambda_{a,b}$ are, indeed, unity.

The bottom line is that a sequence of equally spaced focusing and defocusing lenses of the same strength $1/f$ forms a stable optical system provided that the lenses are separated no further than $2f$. The reader is invited to think of a qualitative physical explanation to the fact that the stability is lost for $L > 2f$.

It turns out that stabilization of a system by applying a periodic sequence of focusing and defocusing interactions is a universal phenomenon making its mark across all areas of physics, from classical mechanics, where periodic jerking of the support of a rigid pendulum may render the otherwise unstable equilibrium point (where the pendulum is pointing up) stable, to trapping charged particles in *radio-frequency traps* (Fig. 5.3) and in *mass spectrometry*, to stabilizing particle trajectories in *alternating-gradient accelerators*, etc. See Prob. 9.6 in the book by Budker et al. (2008) for a tutorial treatment of the *inverted pendulum*.

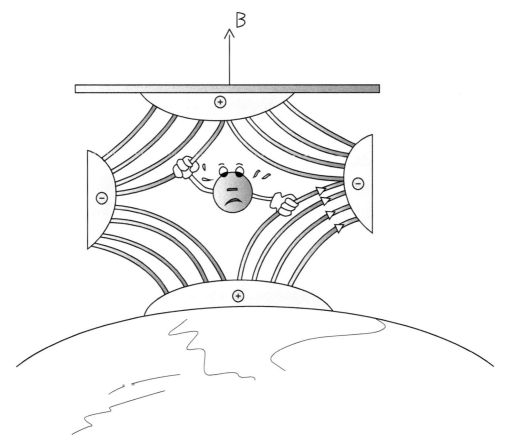

Fig. 5.3 Ion in a trap.

5.3   Nanoparticle optics

Consider a small spherical particle made of a transparent material with refractive index n, for instance, a *nanodiamond* particle with $n \approx 2.4$. The particle is inserted into a free-space plane monochromatic light wave whose wavelength is much greater than the diameter of the particle.

What is the electric field inside the particle?

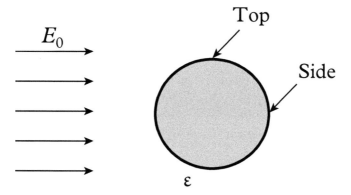

Fig. 5.4 Dielectric sphere in a uniform external field. "Top" and "Side" indicate points where it is particularly convenient to match the boundary conditions.

Solution

Since the wavelength is much larger than the size of the particle, we are in the *electrostatic limit*, where the time dependence of the field inside the particle follows that in the light wave, and where the field of the light wave can be considered uniform and quasi-static.

Assuming nonmagnetic particle ($\mu = 1$), we have a problem of a *dielectric sphere* with dielectric constant $\varepsilon = n^2$ in a uniform field. We encounter similar problems in a variety of physical contexts, see, for example Problems 2.1 and 6.6. The external field $\mathbf{E}_0$ polarizes the particle inducing a dipole moment $\mathbf{d}$. The field inside the particle, $\mathbf{E}_{in}$, is uniform, and the field outside consists of the field $\mathbf{E}_0$ and the dipole field

$$\mathbf{E}_d(\mathbf{r}) = \frac{-\mathbf{d} + 3(\mathbf{d} \cdot \hat{r})\hat{r}}{r^3}. \tag{5.15}$$

Here, $\mathbf{r}$ is the vector from the center of the sphere, and $\hat{r}$ is the corresponding unit vector.

To find $\mathbf{E}_{in}$, we match the boundary conditions. At the "Top" of the sphere (Fig. 5.4), we use the continuity of the tangential component of $\mathbf{E}$:

$$\mathrm{E}_0 - \frac{d}{a^3} = \mathrm{E}_{in}. \tag{5.16}$$

Here a is the radius of the sphere.

On the "Side," we match the electric field outside the sphere and the induction $\varepsilon \mathrm{E}_{in}$ inside, since the normal component of induction must be continuous:

$$\mathrm{E}_0 + \frac{2d}{a^3} = \varepsilon \mathrm{E}_{in}. \tag{5.17}$$

Combining the two conditions, we arrive at the sought answer

$$\boxed{\mathrm{E}_{in} = \frac{3\mathrm{E}_0}{2 + \varepsilon} = \frac{3\mathrm{E}_0}{2 + n^2}.} \tag{5.18}$$

As it should be, $\mathrm{E}_{in} = \mathrm{E}_0$ when $\varepsilon = 1$. For $\varepsilon > 1$, the external field is reduced inside a dielectric sphere.

5.4   Diffraction angle

If a beam of light of wavelength λ is confined in a transverse direction to a characteristic size D, the light acquires divergence with a characteristic *diffraction angle* of $\theta = \lambda/D$ (assuming $\lambda/D \ll 1$).

Derive this result straight from the Maxwell's equations (and in one line). Assume for simplicity (so that the Maxwell's equations for empty space are used) that the confinement is accomplished by focusing a plane wave, for instance, with a *cylindrical lens*.[1]

---

[1]This problem was suggested by Prof. Max Zolotorev.

Solution

Let us assume for definiteness that light propagates along $-\hat{z}$, it is linearly polarized along $\hat{x}$, and that the light beam is of finite extent in the $\hat{y}$ direction. Let us take a circulation loop of dimensions $D \times L$ as shown in Fig. 5.5 that is in the x-y plane. Ignoring numerical factors of order unity, we can estimate

$$\oint \mathbf{E} \cdot d\ell \approx \mathrm{E}L, \tag{5.19}$$

where $\mathbf{E}$ is the electric field of the light. According to the Maxwell's equation for the curl of electric field, the circulation should be proportional to the time derivative of magnetic flux through the loop. But wait a minute! Magnetic field should be perpendicular to both $\mathbf{E}$ and the propagation direction, so the flux through our loop is nominally zero. How can this be?

Of course, what happens here is that, due to the spatial confinement, the beam becomes divergent in the $\hat{y}$ direction with characteristic angle θ, and we can write

$$\oint \mathbf{E} \cdot d\ell \approx \mathrm{E}L \approx \frac{\omega}{c} \mathrm{B}LD\theta, \tag{5.20}$$

where ω is the frequency of the light, which comes from the time derivative; $\mathrm{B} = \mathrm{E}$ is the magnitude of the light magnetic field, and $\mathrm{B}LD\theta$ is an estimate for the magnetic flux through the circulation loop. Taking into account that $\omega/c = 2\pi/\lambda$, we arrive at the sought for estimate of the diffraction angle (modulo the factor of 2π which is OK as we have not been tracking numerical factors).

The assumption that the light is x-polarized is not restrictive. If we assume that the light is y-polarized, we can arrive at an identical diffraction angle using the circulation of magnetic rather than electric field. It follows that it is true for any polarization.

The present derivation is curious in that, while diffraction is usually obtained using *Fourier transform*, here the same result is obtained with a rather different approach.

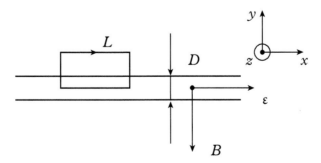

Fig. 5.5 Derivation of the diffraction angle for a beam of x-polarized light propagating along $-\hat{z}$ and spatially confined in the y direction to the size D.

5.5   Diffraction on an edge

A plane monochromatic light wave of wavelength λ falls on a thin screen, and the light diffracts on the edge (Fig. 5.6, top). If we observe the *diffraction pattern* a distance L from the screen, we will find that the intensity as a function of the transverse coordinate x is as shown in Fig. 5.6, bottom.

Without performing any involved calculations, estimate the coordinate x_0 corresponding to the first maximum of the oscillation pattern, which is also the characteristic spatial scale for the intensity decay for $x < 0$. How does x_0 scale with L and λ?

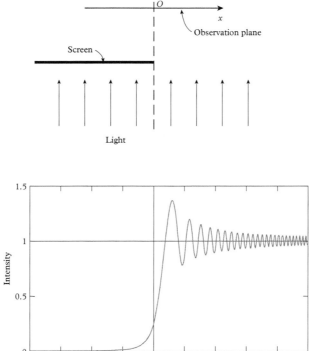

Fig. 5.6 Top: light diffracting on an edge of an opaque screen. Bottom: intensity distribution of monochromatic light in an observation plane downstream from the edge resulting from the diffraction.

Solution

Consider a spot on the observation plane, a transverse distance $-x_0$ from the geometrical edge of the screen. If there were no screen, the wave from the *Huygens source* at the edge of the screen would have some phase shift with respect to the wave that travels directly towards the observation plane. Since we have the screen, we do not have that "direct" Huygens source, so the intensity dies off on the transverse length scale given by the condition that the path difference between the two Huygens sources is $\approx \lambda/2$. The "direct" path is L, while the path from the edge "source" is $\sqrt{L^2 + x_0^2}$. Equating the difference to $\lambda/2$ and expanding for $x_0 \ll L$, we get

$$\boxed{x_0 \approx \sqrt{L\lambda}} \,. \tag{5.21}$$

We can also make an "educated guess" based on *dimensional analysis*. The characteristic distance x_0 should, presumably, increase with both L and λ, and there are no other "letters" in the problem, leaving the expression in Eq. (5.21) a reasonable option with units of distance. (Unfortunately, this is not a unique option because any combination of the form $L^q \lambda^{1-q}$ has dimensions of length.)

Note that, in this case, the characteristic diffraction angle $\theta = x_0/L = \sqrt{\lambda/L}$ depends on L. This is different from the case of diffraction on a finite slit, for which the diffraction angle, $\approx \lambda/D$, where D is the slit size, is independent of L.

5.6   Black-body radiation

What is the root-mean-square (RMS) value of the electric field of the *black-body radiation* at a temperature of $T = 300$ K? (Make a guess based on your physical intuition before calculating.) Assume that the radiation is in equilibrium with the surrounding bodies (for example, think of the black-body radiation in your room).

Solution

According to the *Stefan–Boltzmann law*, the power radiated per unit surface area by a black body with temperature T is

$$j = \sigma T^4, \tag{5.22}$$

where $\sigma \approx 5.67 \times 10^{-5}$ erg/s/cm^2 is the *Stefan–Boltzmann constant*. Now, we can also write the same quantity using the *Poynting vector* as

$$j = \frac{1}{4} c \frac{\mathrm{E}^2}{4\pi}, \tag{5.23}$$

where E^2 is the r.m.s. electric field of the black-body radiation, $\mathrm{E}^2/4\pi$ is the r.m.s. energy per unit volume (including the equal electric and magnetic contributions to the energy), and the factor of $1/4$ takes into account that half of the black-body-radiation photons in equilibrium are moving from the surface, and that the average normal projection of velocity for these is $c/2$ (prove this!).

Equating the two previous expressions, we get

$$\mathrm{E}^2 = \frac{16\pi\sigma T^4}{c} \approx (0.0277 \text{ esu})^2 \approx (8.32 \text{ V/cm})^2, \tag{5.24}$$

where in the last step we substituted 1 esu for 300 V/cm.

Thus the r.m.s. electric field for room-temperature black-body radiation is

$$\boxed{\mathrm{E}_{r.m.s.} \approx 8.32 \text{ V/cm},} \tag{5.25}$$

perhaps an unexpectedly large value (depending on how good a feeling you have for orders of magnitude of physical quantities).

5.7   Laser vs. thermal light source

Imagine we had a fast photodetector that could detect *instantaneous intensity of light*, by which one means intensity averaged over an optical cycle. Suppose we use the detector to measure a light beam from a powerful non-laser light source, for example, an incandescent lightbulb or a discharge lamp. The light beam passes through a polarizer on the way to the detector. Please feel free to assume any values for the other relevant parameters such as, for instance, the average light power on the detector.

(a) What is the most probable reading of the detector?

(b) Same for a laser.

Solution

(a) The light sources discussed in this part are examples of *chaotic light sources*, which can be thought of as collections of independent dipoles, for instance, the mercury atoms in a lamp, that radiate independently. For simplicity, we can assume that all the dipoles radiate at the same nominal frequency (that of an atomic transition), but their phases are independent. The amplitude of the light field at the detector, which, for the sake of simplicity, we assume to be small enough so that the phase differences of the light at different points of the detector can be neglected, is the sum of the amplitudes of the light field from each individual radiating dipole, which can be visualized with an *Argand diagram*, where the complex amplitudes are drawn as vectors on a complex plane (Fig. 5.7a). Note that the two axes on the Argand diagram represent two independent phases of light of a given polarization, and should not be confused with the directions of polarization.

Since the phases of the individual dipoles are random, the total amplitude is a result of a *random walk* starting from the origin, and having individual steps represented by the amplitude vectors of the light from each of the dipoles. The result of such random walk is that the most probable value of the electric field is zero!

Let us make the analysis more precise by invoking the random-walk theory. Let us first consider one phase, for instance, sine. If we have a large number N of the emitters, each one contributing randomly to the sine component E_s with a random "step" E_0, then the probability distribution for finding the sine amplitude between E_s and $E_s + dE_s$ is, up to a normalization factor,

$$p(E_s) \propto \exp\left\{-\frac{E_s^2}{2NE_0^2}\right\} dE_s. \tag{5.26}$$

The combined probability distribution for the independent cosine and sine amplitudes is then

$$p(E_c, E_s) \propto \exp\left\{-\frac{E_s^2 + E_c^2}{2NE_0^2}\right\} dE_s dE_c = \exp\left\{-\frac{I}{\bar{I}}\right\} d\varphi E dE, \tag{5.27}$$

where we have introduced the instantaneous cycle-averaged intensity I, its mean value $\bar{I}$, the *phase angle* φ, and the total field amplitude E, which is non-negative.

We can now convert this into a distribution of cycle-averaged instantaneous intensity by integrating over φ and using the fact that $I \propto E^2$, so, correspondingly, $dI \propto 2E dE$. With this, Eq. (5.27) becomes

$$p(I) \propto \exp\left\{-\frac{I}{\bar{I}}\right\} dI, \tag{5.28}$$

where, again, we have neglected the 2π and other multiplicative constants. The function $p(I)$ is sketched in Fig. 5.7b.

From Eq. (5.28) and Fig. 5.7b, we are led to conclude that the most probable cycle-averaged light intensity we would measure from a polarized chaotic light source is, in fact, $\boxed{zero.}$

Note that the characteristic *correlation time* of the cycle-averaged power is determined by the time the individual dipoles maintain their relative phase. This is the same time scale that dictates the *spectral width* of the radiation.

A reader who has been paying attention is probably wondering, why is it important to specify that the light is polarized? As it turns out, because for unpolarized light, we need to consider two independent polarizations, the extra degrees of freedom in the random-walk problem result in a significant modification of the distribution of instantaneous intensities [see, for example, Sec. 13.3.1 in the book by Mandel and Wolf (1995)]. In fact, the probability distribution goes to zero at zero intensity in this case instead of being maximal as in the case of polarized light. A curios consequence of this is that, apparently, it is possible to tell if chaotic light is polarized or not by measuring the distribution of light-intensity readings from a fast polarization independent detector.

(b) With a laser, in which the individual dipoles radiate *coherently*, i.e., with correlated phases, the situation is radically different. Let us assume a *single-mode laser*, in which the emission is nominally at just one frequency. Adding the amplitudes of the radiation from the individual dipoles, which now all point in the same direction, we arrive at some finite total amplitude of the light field. If, for some reason, a dipole emits with

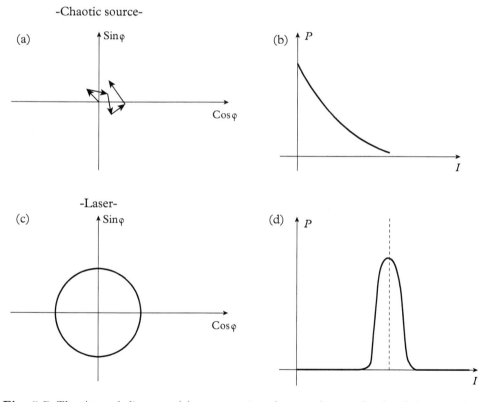

Fig. 5.7 The Argand diagram, (a), representing the complex amplitude of the light from a chaotic source of a particular polarization. The electric-field contributions of individual radiating dipoles are shown with arrows, and the result is analogous to what happens in random walk. In (b), the probability distribution of instantaneous cycle-averaged intensity is sketched for a beam of polarized chaotic light. The analogous plots for the case of a single-mode laser are shown in (c) and (d).

a "wrong" phase (for example, due to *spontaneous emission* rather that *stimulated emission*), this leads to a small rotation of the overall amplitude, and, eventually, to *phase diffusion* around a circle on the Argand diagram, as sketched in Fig. 5.7c. In this case, the most probable amplitude (Fig. 5.7d) and cycle-averaged power are both non-zero. In fact, the most probable detected instantaneous cycle-averaged intensity is just the average intensity on the detector.

The topics briefly touched upon in this Problem are discussed in detail in our favorite book on *The Quantum Theory of Light* by Loudon (2000).

5.8 Correlation functions for light and Bose condensates

Chaotic light produced by an ensemble of uncorrelated emitters has rather remarkable properties that we touched upon in Prob. 5.7 and continue to explore here. We adopt the following model for the light source (Loudon, 2000): the source consists of a large ensemble of atoms, each of which radiates continuously, at the same frequency and the same amplitude, but with a random phase. It is assumed that the phase of the radiation for each atom remains constant for some time, and changes randomly with a characteristic time τ. The light detector and the source are arranged so that the light electric fields from each atom add coherently at the detector, which is assumed to measure the *instantaneous intensity* of the light.

According to the model of the chaotic light source, the electric field on the detector is a sum of electric fields from individual atoms, which contribute the fields of the same amplitude and frequency, but with a random phase ϕ_j, where $1 \leq j \leq N$, and N is the total number of atoms. The corresponding phase factors $\exp(i\phi_j)$ can be represented as unit-length vectors on a complex plane with the direction given by ϕ_j (an *Argand diagram*). The overall amplitude of the light at any moment can be thought of as a result of two-dimensional *random walk*, for which the most probable result is zero. In other words, starting a random walk at the origin, the most probable location after N steps is still the origin. The most probable intensity for polarized light is, correspondingly, also zero (Prob. 5.7)!

A key property of a chaotic light source is that the field from different atoms in the source cancel each other, and, in fact all that we see is the fluctuations around the perfect cancellation. We note that this explains why the intensity of a chaotic source is proportional to N, while for a *coherent source* such as, for example, a laser, the intensity scales as the square of the number of the emitters.

(a) Coherence properties of light are frequently characterized by r^{th}-order *correlation functions* (Loudon, 2000) that relate the values of the light electric field and its higher powers with the corresponding values at a later time. For example, the first-order correlation function is

$$g^{(1)}(\tau) = \frac{\langle \mathrm{E}^*(t+\tau)\mathrm{E}(t)\rangle}{\langle \mathrm{E}^*(t)\mathrm{E}(t)\rangle}, \tag{5.29}$$

where the averaging denoted by $\langle \ldots \rangle$ is carried out over all times t. The expression in the denominator is just the average light intensity. For chaotic light, we expect $g^{(1)}(-\infty) = g^{(1)}(\infty) = 0$ because of the finite *phase memory* of the light. On the other hand, $g^{(1)}(0) = 1$ because the same nonzero expression (i.e., the average light intensity, $\langle I \rangle$) appears in the numerator and the denominator.

The second-order correlation function is

$$g^{(2)}(\tau) = \frac{\langle \mathrm{E}^*(t+\tau)\mathrm{E}(t+\tau)\mathrm{E}^*(t)\mathrm{E}(t)\rangle}{\langle I \rangle^2}. \tag{5.30}$$

Using the model of the chaotic light source described previously, derive the value of the second-order correlation function for zero delay, $g^{(2)}(0)$.

(b) Derive the value of $g^{(r)}(0)$ for an arbitrary natural value of r, where the rth-order zero-time-delay correlation function is defined as

$$g^{(r)}(0) = \frac{\langle [\mathrm{E}^*(t)\mathrm{E}(t)]^r \rangle}{\langle I \rangle^r}. \tag{5.31}$$

Solution

(a) For a chaotic source, the average intensity of the radiation is N times the intensity produced by an individual atom. This takes care of the evaluation of the denominator in Eq. (5.30), which comes to N^2 times the square of the intensity produced by one atom. The numerator of (5.30), explicitly written in terms of the electric fields of individual atoms, contains the sum of the terms

$$e^{i(-\phi_j+\phi_{j'}-\phi_{j''}+\phi_{j'''})}, \tag{5.32}$$

where each of the indices $j,\ldots,j'''$ runs between 1 and N, and each of the phases is random. It is clear that only terms where the overall phase is zero will survive time averaging.

One possibility for the overall phase to be zero is for each of the indices to be the same. There are N such terms.

The other possibilities for the overall phase to vanish are: $j = j'$, $j'' = j'''$, $j' \neq j''$, and $j = j'''$, $j' = j''$, $j \neq j'$. In the first of these two cases, we have N ways to pick j and $N-1$ ways to pick j', so that we have $2N(N-1)$ terms of this type in all, which is a vastly larger number than N for the case of many atoms.

Combining the numerator and denominator in Eq. (5.30) with $\tau = 0$, we find

$$\boxed{g^{(2)}(0) = 2 - 1/N,} \tag{5.33}$$

which tends to 2 for large N.

This result demonstrates *photon bunching* for a chaotic light source. It is also the essence of the famous *Brown–Twiss effect*, which is the basis of a technique called *intensity interferometry* (Brown, 1974).

> There are circumstances in which the simplified classical model of emitters, adopted in this problem, fails. For example, for a single emitter, $g^{(2)}(0) = 0$ (*photon anti-bunching*), a property that is widely used to identify such *single-quantum emitters*.

(b) We can solve the problem in analogy with Part (a). Nonzero contributions in the numerator of the rth-order correlation function come from the terms where a phase $\phi_j + \phi_{j'} + \ldots \phi_{j^{r}}$ is multiplied by its complex conjugate. Adopting the large-N approximation, we have approximately N ways to pick each of the indices, which has to be further multiplied by the number of their permutations ($r!$), giving the total number of terms as $\approx r! N^r$, and the final result as

$$\boxed{g^{(r)}(0) \approx \frac{r! N^r}{N^r} = r! \, .} \tag{5.34}$$

> It is interesting to note that the increase of the second- and higher-order correlation function at zero delay is observed not only for light, but also for a system of a rather different nature—an atomic *Bose–Einstein condensate* (BEC) (Dall et al., 2011). Here, instead of detecting light intensity (or counting photons), one is detecting BEC atoms, and measuring their spatial correlations.

5.9 Pulsed laser repetition rate

Consider a pulsed laser continuously producing a train of short light pulses with a *pulse repetition rate* of, say, f_{rep}=250 MHz.

(a) What is the spectrum of the laser light?

(b) Suppose a certain application requires a train of pulses with a higher repetition rate, e.g., four times higher or 1 GHz for our example.

Is it possible to boost the repetition using only passive optical components (such as, for example, mirrors) external to the laser? Please describe how this might be done.

Solution

(a) The short duration of a pulse means that the spectrum should be broad, with the minimal spectral span $\Delta\nu$ determined by the uncertainty relation

$$\Delta\nu\Delta t \approx \frac{1}{2\pi}, \tag{5.35}$$

where Δt is the duration of one pulse. The periodic pulsing imposes a structure on the overall broad *spectral envelope* determined by the *Fourier transform* of an individual pulse. Let us consider which frequency components may be present in the spectrum. Note that any frequency component $N \cdot f_{rep}$ (N is an integer) that is *phase locked* with the fundamental frequency f_{rep} (so that the fields corresponding to these harmonics are all maximal at the time of a pulse) will have a maximum at the times of the consequent pulses, while at any other point in time, the contribution of such waves sums up to zero with an accuracy that increases with the number of components that we add (Fig. 5.8). Thus, we conclude that the spectrum of our laser is a *frequency comb* with the separation between adjacent *frequency-comb teeth* equal to f_{rep}.

We note that a temporal comb produces a frequency comb (and vice versa). Thus, the function that represents a comb, the *shah function* (so named after a letter in the Russian alphabet that looks like a three-tooth comb), has the property that it goes into itself upon Fourier transform. Another well-known function that has the same property is a Gaussian.

(b) To achieve the desired multiplication of the repetition rate, one can employ *spectral filtering*, for example, with a *Fabry–Perot* cavity that can be constructed with two mirrors. The frequency separation of the transmission peaks of the cavity should be chosen so that it is four times (in our particular example) the repetition rate. When the cavity is adjusted to transmit every fourth tooth of the original frequency comb, the output light power will be four times lower (assuming a perfectly adjusted Fabry–Perot cavity with lossless mirrors), but the output repetition rate will be 1 GHz, as desired. The output pulse rate is the *round-trip period* in the cavity. Note that the energy in a single temporal pulse in the new beam is 16 times smaller than that in the original beam.

> Spectral filtering of frequency-comb lasers finds application in devices called *astro-combs* (see Fig. 5.9) that are used for precise calibration of astrophysical spectrometers, for accurate measurements of the speeds of astronomical bodies, and for the search for *exoplanets* that fascinate one's imagination due to the glimpse of a possibility that some of them might harbor *extraterrestrial life*.

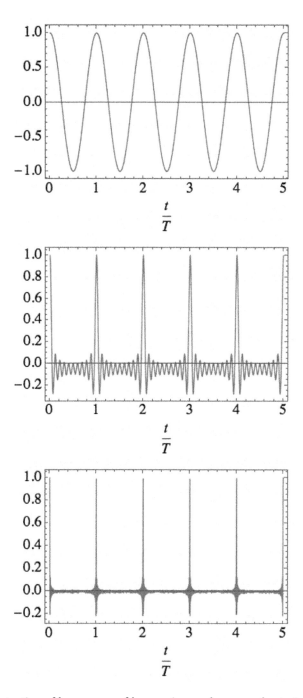

Fig. 5.8 An illustration of how a sum of harmonics produces a pulse train with a repetition rate equal to the fundamental frequency $f_{rep} = 1/T$. Top: a cosine wave; middle: a normalized sum of 10 harmonics $cos(2\pi Nt/T)$, where $N = 1, \ldots, 10$ is the harmonic number, T is the period of the fundamental wave; bottom: a similar sum of 100 such harmonics.

Fig. 5.9 This photo taken in 2008 at the Smithsonian Astrophysical Observatory shows Prof. Ronald Walsworth (right) and Dr. Chih-Hao Li (who had obtained his Ph.D. from Berkeley in 2005) with their *astro-comb*, a device utilizing spectral filtering along the lines of our discussion in this problem. Photo by Jon Chase/Harvard University.

5.10 Beamsplitter

There is hardly an element that is more essential for understanding optical processes than a *beamsplitter*, as we will illustrate here with two examples.

(a) Consider an uncoated glass plate that can be used as a partial reflector. If a beam of light is directed at a plate as shown in Fig. 5.10(a), for near-normal incidence, there is an amplitude of about 0.2 in magnitude for common glass for reflection from each of the two surfaces. If the two reflected beams overlap, they interfere and the total reflected intensity can be anywhere from zero to about 16% depending on the relative phase of the two reflections, in turn dependent on the thickness of the plate δ.

What is the total reflection intensity in the limit $\delta \to 0$, in the approximation that the refection amplitudes are small?

(b) Now consider a 50-50 beamsplitter that is designed to eliminate the reflection from one of the surfaces, which is typically accomplished with *antireflection dielectric coating*, Fig. 5.10(b). Suppose further that two identical photons simultaneously impinge on the beamsplitter from ports A and B with one photon coming into each of the ports.

Describe what one will see at the output (ports C and D).

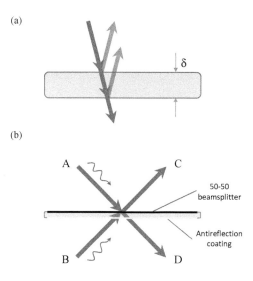

Fig. 5.10 (a) Reflection from an uncoated glass plate. (b) A beamsplitter consisting of a glass plate with one surface coated for partial (50%) reflection and the other surface coated to avoid any reflection. Here we consider a situation where a pair of photons is sent on the beamsplitter through "ports" A and B.

Solution

(a) As δ, or more precisely, $n\delta$ (where $n \approx 1.5$ is the *refractive index* of glass) becomes negligibly small compared to optical wavelength, the path-length difference for the two reflected beams vanishes along with the corresponding phase difference. However, in comparing the relative phases of these two reflections, it is essential to remember that when a reflection occurs at a boundary with a medium with a higher refractive index, there is a phase shift by π, while this phase shift is absent when the *reflection* is from a medium with lower refractive index. We conclude that the total reflection amplitude vanishes for $\delta \to 0$.

(b) We first note that it is not a trivial matter to produce a correlated two-photon initial state considered in this problem, but this feat is nevertheless routinely accomplished with modern *nonlinear-optics* techniques.

What are the possible outcomes of the experiment we have set up here? In principle, we can end up with two photons in the "output port" C (call this CC), two photons in D (DD), and one photon in each of the ports C and D (CD and DC, though there is no way to distinguish between these latter two).

Let us consider the quantum-mechanical amplitudes for these different possibilities. There is only one possible amplitude for, say, the CC outcome, that for both initial photons to go to C. This amplitude is finite and the same holds for the DD outcome.

For the CD/DC outcome, there are two possibilities that we can schematically designate (A→D,B→C) and (A→C,B→D), whose amplitudes we need to add. In the first of these possibilities, both photons go through the beamsplitter without reflection, while in the second possibility, we have both photons undergoing reflections. Note that, in the latter case, one of these reflections (A→C) comes with a π phase shift discussed in part (a), while for the other (B→D) such phase shift is absent, so there is an overall minus sign for the amplitude with reflections. This leads us to a (possibly astonishing) conclusion that the total amplitude for CD/DC vanishes, and both photons always go to the same port! This is known as the *Hong-Ou-Mandel effect*.

We remark that the amplitudes of the AB→CC and AB→DD processes are larger than one would obtain for independent photons. This is due to *quantum interference*.

To avoid possible confusion, we caution the reader that in quantum-optics literature, a more symmetric model of a beamsplitter is usually considered where, for each transmission/reflection there is a $\pm\pi/2$ phase shift between the transmitted/reflected beam. The phase choices are nicely explained in Sec. 3.2 of our beloved book by Loudon (2000).

5.11 Rotating linear polarization

Linearly polarized light can be decomposed into two counter-rotating circular polarizations. Now suppose that the linearly polarized light passes through a $\lambda/2$ plate that is rotating around its axis with frequency Ω.

Describe the polarization state of light after the $\lambda/2$ plate. Show that this polarization state can be represented as a sum of the counter-rotating circular polarizations with unequal frequencies. What are these frequencies?

Solution

If the initial polarization is linear along the x-axis, we can write the electric field of the light beam:

$$\mathbf{E}_i = \hat{x} E_0 \cos \omega t, \tag{5.36}$$

where E_0 is the field amplitude and ω is the frequency of the light. A rotating $\lambda/2$ plate makes an angle Ωt with the x-axis at time t. It rotates the linear polarization by twice the angle between the plate axis and the polarization axis. Therefore, the electric field of the light bean after the plate is

$$\mathbf{E}_f = (\hat{x} \cos 2\Omega t + \hat{y} \sin 2\Omega t) E_0 \cos \omega t, \tag{5.37}$$

representing linearly polarized light whose polarization rotates with frequency 2Ω.

In order to write this as a sum of circular polarizations, we use trigonometric identities for product of cosine and sine functions and obtain:

$$\mathbf{E}_f = \frac{E_0}{2} \left[(\hat{x} \cos(\omega + 2\Omega)t + \hat{y} \sin(\omega + 2\Omega)t) + \right.$$
$$\left. + (\hat{x} \cos(\omega - 2\Omega)t - \hat{y} \sin(\omega - 2\Omega)t) \right]. \tag{5.38}$$

Therefore, the light beam is a sum of a right-hand circularly polarized sideband at frequency $\omega + 2\Omega$ and a left-hand circularly polarized sideband at frequency $\omega - 2\Omega$. A hand-waving representation of the two circular polarization components is shown in Fig. 5.11.

Fig. 5.11 Hand-waving explanation of optical polarization rotation, with arms representing the direction of electric field. Top: the two circular polarization components rotate at the same frequency, adding up to a vertical linear polarization. Bottom: the two circular polarization components rotate at slightly different frequencies, adding up to a rotating linear polarization.

6

Quantum, Atomic, and Molecular Physics

Two fermions walk into a bar. The
first says "I'd like a vodka martini
with a twist." The second says
"Dammit, that's what I wanted!"

Physics on Your Feet: Berkeley Graduate Exam Questions: or Ninety Minutes of Shame but a PhD for the Rest of Your Life! Dmitry Budker and Alexander O. Sushkov, Oxford University press. © Dmitry Budker, Alexander O. Sushkov, Vasiliki Demas 2015, 2021. DOI: 10.1093/oso/9780198842361.003.0006

6.1 Magnetic decoupling of spins

Consider a system of two nonequivalent spins (both spin-1/2), $\mathbf{I}$ and $\mathbf{S}$, with associated magnetic moments $\boldsymbol{\mu}_I = \gamma_I \mathbf{I}$ and $\boldsymbol{\mu}_S = \gamma_S \mathbf{S}$, and the interaction between the spins described by the Hamiltonian

$$H = J\mathbf{I} \cdot \mathbf{S}, \tag{6.1}$$

where J is a constant. An example of such a system is a diatomic molecule with all electrons paired, and the two spins being those of the two nuclei. The interaction described by Eq. (6.1) is historically called *J-coupling*, and is different from the usual dipole-dipole interaction because the latter vanishes upon the averaging over the molecular orientation, and J-coupling, being mediated by the molecular electrons, does not.

(a) What are the energy levels of the system in the absence of external fields? What are the *degeneracies* of these levels?

(b) Sketch the evolution of the *eigenenergies* of the system as a function of an applied static magnetic field.

Solution

The problem of finding the eigenstates and eigenenergies of a system with J-coupling in the presence of a magnetic field is exactly analogous to that of the *Zeeman effect* in a manifold of hyperfine atomic levels (see, for example, Auzinsh et al. 2010, Sec. 4.2).

(a) At zero magnetic field, we have a *triplet* (total angular momentum $F=1$) with energy $+J/4$, and a *singlet* ($F=0$) with energy $-3J/4$ (Fig. 6.1).

Using this *coupled basis* of $|F, M\rangle$ states, we can write the Hamiltonian including J-coupling and the Zeeman interaction as

$$H = \begin{pmatrix} \frac{J}{4} - (\mu_I + \mu_S)\mathrm{B} & 0 & 0 & 0 \\ 0 & \frac{J}{4} + (\mu_I + \mu_S)\mathrm{B} & 0 & 0 \\ 0 & 0 & \frac{J}{4} & -(\mu_I - \mu_S)\mathrm{B} \\ 0 & 0 & -(\mu_I - \mu_S)\mathrm{B} & -\frac{3J}{4} \end{pmatrix}, \qquad (6.2)$$

where we order the states $(|1, 1\rangle, |1, -1\rangle, |1, 0\rangle, |0, 0\rangle)$.

(b) At small magnetic fields, the $F = 1$, $M = 0$ (M is the projection of the total angular momentum on the direction of the magnetic field) and $F = 0$ experience a shift that is quadratic in the magnetic field. At higher fields, where the Zeeman shifts exceed the zero-field J-coupling, the shift of these states becomes linear in the magnetic field. The $M = \pm 1$ states experience linear shift at all values of the magnetic field. In the high-field limit, the eigenstates correspond to the states of the *decoupled basis*, where the states are characterized by the fixed projections of the individual spins of each of the nuclei on the magnetic field. The diagram on Fig. 6.1 is a direct analog of the *Breit–Rabi* diagram in atomic physics, which describes *magnetic decoupling* of the hyperfine structure.

A notable feature seen in Fig. 6.1 is the *level crossings*: that of the triplet states at zero field, and those of the two lowest finite-field levels. These level crossings can be contrasted to the *level anti-crossing* of the $M = 0$ states. Finite off-diagonal interactions turn level crossings into anti-crossings (a.k.a. *avoided crossings*), a rather ubiquitous situation in many physical problems, see, for instance, Prob. 6.2.

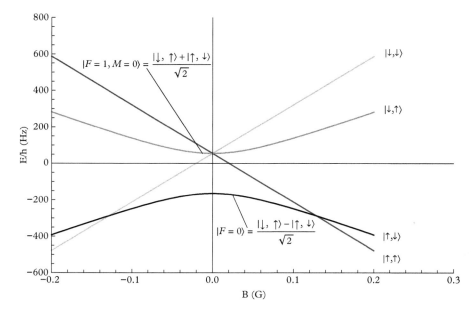

Fig. 6.1 Energy levels corresponding to the full Hamiltonian [Eq. (6.2)] including J-coupling and the interaction with external magnetic field. The numerical factors are chosen to correspond to realistic parameters for a molecule with a ^{13}C nucleus in the vicinity of a proton. At low fields, the eigenstates correspond to the coupled basis with well-defined values of F and M. At high fields, individual projections of the two spins (shown with vertical arrows) are well-defined. The zero-field splitting of the two $M = 0$ levels can be thought of as *level anti-crossing* due to the J-coupling interaction.

Magnetic decoupling illustrated as "detripling"...

6.2 Level anticrossing

Consider a system of two energy levels with an energy gap Δ. In the absence of any coupling between the two states, by setting the energy of the one of the states to be $E_1 = 0$, their Hamiltonian can be written as

$$H = \begin{pmatrix} 0 & 0 \\ 0 & \Delta \end{pmatrix}. \tag{6.3}$$

The energy spectrum of the system as a function of Δ is sketched in Fig. 6.2a. At $\Delta = 0$, there is a *level crossing*.

Now, suppose that there is some *off-diagonal perturbation* δ that couples the two states. In general, δ is complex. In this case, the Hamiltonian becomes

$$H = \begin{pmatrix} 0 & \delta \\ \delta^* & \Delta \end{pmatrix}, \tag{6.4}$$

and, instead of the level crossing at $\Delta = 0$, we have a *level anti-crossing* as sketched in Fig. 6.2b.

The situation becomes slightly less obvious when the levels are unstable. This can be described by a *non-Hermitian Hamiltonian*

$$H = \begin{pmatrix} -i\Gamma_1/2 & \delta \\ \delta^* & \Delta - i\Gamma_2/2 \end{pmatrix}, \tag{6.5}$$

and the level widths $\Gamma_{1,2}$ are shown graphically as the finite widths of the lines representing the two levels (Fig. 6.2c).

The question is: is there an anticrossing for states of finite width? Formulate the criterion for what should be called "anticrossing" in this situation, and find the conditions for such anticrossing to occur.

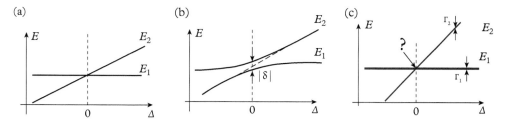

Fig. 6.2 Level crossing (a) for states without off-diagonal coupling vs. level anticrossing (b) for states with nonzero off-diagonal coupling. In the latter case, the situation is more subtle if the states have finite widths (c).

Solution

For the case of levels of finite width, it is reasonable to define level anticrossing by the existence of a splitting of the real parts of the complex energies of the *eigenstates* at all values of Δ. Let us find the complex *eigenenergies* of the Hamiltonian of Eq. (6.5) by solving the *characteristic equation*

$$\begin{vmatrix} -i\Gamma_1/2 - \lambda & \delta \\ \delta^* & \Delta - i\Gamma_2/2 - \lambda \end{vmatrix} = 0, \tag{6.6}$$

which reduces to

$$\lambda_{1,2} = \frac{\Delta - \frac{i}{2}(\Gamma_1 + \Gamma_2) \pm \sqrt{\left[\Delta - \frac{i}{2}(\Gamma_2 - \Gamma_1)\right]^2 + 4|\delta|^2}}{2}. \tag{6.7}$$

It is interesting to note that when the two levels have the same width ($\Gamma_1 = \Gamma_2$), then the quantity under the square root in Eq. (6.7) is real and positive, and there is always an anticrossing.

Suppose now that the two widths are different ($\Gamma_1 \neq \Gamma_2$). For $\Delta \neq 0$, the quantity under the square root is complex, and the real parts of the two complex eigenenergies are distinct. Finally, consider the case of $\Delta = 0$ (which we are most interested in). The quantity under the square root is now real, however, it can be either positive (anticrossing) or negative. In the latter case, the square root is pure imaginary, and the real parts of the eigenenergies coincide, so the anticrossing is absent. The condition for the existence of the anticrossing is thus

$$\boxed{|\delta| > |\Gamma_2 - \Gamma_1|/4.} \tag{6.8}$$

This result was derived by V. V. Sokolov and V. G. Zelevinsky and is also discussed in the book by Khriplovich (1991) (Sec. 9.5).

6.3 Bound states in a potential well

A potential well is described by $V = -V_0$ for $0 \leq r \leq R$ and $V = 0$ elsewhere. Is there always a bound state in this potential for a particle of mass m? Please discuss this for:

(a) one-dimensional case;

(b) three-dimensional case;

(c) two-dimensional case.

The authors found two ways to solve this problem (although there are probably more). The first solution given here is approximate and applicable to a potential well of any shape, whereas the second solution is applicable to a square-well potential, and makes use of mathematical facts and special functions listed in the following. We leave it up to the reader to choose his or her favorite solution, or come up with their own.

The reader may find the following mathematical facts helpful (or not).

- The radial part of the *Laplacian in cylindrical coordinates* is

$$\frac{1}{r}\frac{\partial}{\partial r}\left(r\frac{\partial}{\partial r}\right) = \frac{\partial^2}{\partial r^2} + \frac{1}{r}\frac{\partial}{\partial r}. \tag{6.9}$$

- The solution of the equation

$$\frac{\partial^2 \psi}{\partial r^2} + \frac{1}{r}\frac{\partial \psi}{\partial r} + k^2 \psi = 0 \tag{6.10}$$

for real k is generally given by

$$\psi(r) = C_1 J_0(kr) + C_2 Y_0(kr), \tag{6.11}$$

where C_i are constants, $J_0(z)$ is the *Bessel function of the first kind*, and $Y_0(z)$ is the *Neumann's (Weber's) Bessel function of the second kind*. The function $Y_0(z)$ diverges at the origin. The asymptotic behavior of J_0 is

$$J_0(z) \approx 1 - \frac{z^4}{4}, \quad |z| \ll 1. \tag{6.12}$$

- The solution of the equation

$$\frac{\partial^2 \psi}{\partial r^2} + \frac{1}{r}\frac{\partial \psi}{\partial r} - \kappa^2 \psi = 0 \tag{6.13}$$

(for real positive κ) that vanishes at infinity is

$$\psi(r) = C_3 \left[J_0(i\kappa r) + i Y_0(i\kappa r) \right] = C_3 H_0^{(1)}(i\kappa r), \tag{6.14}$$

where $H_0^{(1)}$ is the *Hankel function of the first kind*. The asymptotic of this function at small values of κr is $H_0^{(1)} \propto \ln(\kappa r)$.

Solution

If we assume that the particle is localized to a spatial region with linear dimensions of $\approx R$, then we can formulate the following criterion for the existence of a bound state. The total energy of a bound state is negative and is a sum of the potential energy $\approx -V_0$ and the minimum kinetic energy of $\approx \hbar^2/(2mR^2)$ that comes from the localization due to the position-momentum uncertainty. For the total energy to be negative, we need (neglecting numerical factors):

$$\frac{\hbar^2}{mR^2} \lesssim V_0. \tag{6.15}$$

Unfortunately, the argument is not universal. The reason for this is that the particle does not necessarily need to be localized within the potential. A quintessential example of this is a bound state associated with a delta-function potential, where the entirety of the particle's wave function resides outside of the potential.

Hence, let us get to work and solve the problem!

Consider the Schrodinger equation

$$\left[\frac{\hbar^2 p^2}{2m} + U(\mathbf{r})\right]\psi(\mathbf{r}) = E\psi(\mathbf{r}), \tag{6.16}$$

where momentum p is in units of $\hbar$. Since we are interested in solutions of this equation for various forms of the potential $U(\mathbf{r})$ (see the following), it is convenient to work in *momentum representation*, where p is a number rather than an operator:

$$\left[\frac{\hbar^2 p^2}{2m} - E\right]\psi_\mathbf{p} = -\int U_{\mathbf{p}-\mathbf{q}}\psi_\mathbf{q}\frac{d^D\mathbf{q}}{(2\pi)^D}, \tag{6.17}$$

where D is the number of dimensions, and the Fourier transforms of the potential are given by:

$$U_\mathbf{k} = \int U(\mathbf{r})e^{-i\mathbf{k}\cdot\mathbf{r}}d^D\mathbf{r}, \tag{6.18}$$

$$U(\mathbf{r}) = \int U_\mathbf{k}e^{i\mathbf{k}\cdot\mathbf{r}}\frac{d^D\mathbf{k}}{(2\pi)^D}. \tag{6.19}$$

In order to obtain Eq. (6.17), the wavefunction $\psi(\mathbf{r})$ in Eq. (6.16) is expressed in terms of its *Fourier transform* $\psi_\mathbf{q}$, the equation is multiplied by $e^{-i\mathbf{p}\cdot\mathbf{r}}$, and integrated over $\mathbf{r}$:

$$\iint\left[\frac{\hbar^2 p^2}{2m} - E\right]\psi_\mathbf{q}e^{i(\mathbf{q}-\mathbf{p})\cdot\mathbf{r}}d^D\mathbf{r}\frac{d^D\mathbf{q}}{(2\pi)^D} = -\iint U(\mathbf{r})\psi_\mathbf{q}e^{i(\mathbf{q}-\mathbf{p})\cdot\mathbf{r}}d^D\mathbf{r}\frac{d^D\mathbf{q}}{(2\pi)^D}. \tag{6.20}$$

The spatial integral on the left-hand side gives $\delta(\mathbf{q} - \mathbf{p})$, and the momentum integral then gives $\psi_\mathbf{p}$, whereas the spatial integral on the right-hand side is the Fourier transform $U_{\mathbf{p}-\mathbf{q}}$, resulting in Eq. (6.17).

The existence of a weakly bound state is independent of the exact shape of the potential $U(r)$, because the wavefunction of such a state has a spatial extent that is much larger than the extent of the potential R. Therefore, we use an approximation for the Fourier components $U_{\mathbf{k}}$. Since $U(r) = 0$ for $r > R$, we can set the upper limit of the integral in Eq. (6.18) to R. When $|\mathbf{k}| \gg 1/R$, the exponential is rapidly oscillating over the entire range of the integral, and $U_{\mathbf{k}} \approx 0$, whereas when $|\mathbf{k}| \ll 1/R$, we can replace the exponential by unity, and $U_{\mathbf{k}} \approx -g$, where $g = V_0 R$ in one dimension, $g = V_0 \pi R^2$ in two dimensions, and $g = V_0 4\pi R^3/3$ in three dimensions. This is equivalent to approximating the potential by the delta function: $U(\mathbf{r}) = -g\delta(\mathbf{r})$, where g is a parameter proportional to the depth of the potential V_0.

Thus, we make the approximation $U_{\mathbf{k}} = -g$ for $|\mathbf{k}| < \Lambda$, and $U_{\mathbf{k}} = 0$ for $|\mathbf{k}| > \Lambda$, where $\Lambda \approx 1/R$ is the parameter whose value depends on the "shape" of the potential $U(r)$. Substituting this into Eq. (6.17), we let the integral over q run from 0 to Λ, and:

$$\left[\frac{\hbar^2 p^2}{2m} - E\right]\psi_{\mathbf{p}} = g \int \psi_{\mathbf{q}} \frac{d^D \mathbf{q}}{(2\pi)^D} = C, \qquad (6.21)$$

where C is a constant, independent of momentum $\mathbf{p}$. Since p is a number and not an operator, we deduce that

$$\psi_{\mathbf{p}} = \frac{C}{\hbar^2 p^2/2m - E}, \qquad (6.22)$$

and substitute back into Eq. (6.17) to obtain

$$g \int \frac{C}{\hbar^2 q^2/2m - E} \frac{d^D \mathbf{q}}{(2\pi)^D} = C, \qquad (6.23)$$

and finally divide through by C and introduce the magnitude of the bound-state energy $\epsilon = |E| = -E$:

$$g \int \frac{1}{\hbar^2 q^2/2m + \epsilon} \frac{d^D \mathbf{q}}{(2\pi)^D} = 1. \qquad (6.24)$$

Our task is to find out whether Eq. (6.24) has a solution for ϵ for an arbitrarily small binding g.

(a) In one dimension, the integral turns into:

$$g \int_0^\Lambda \frac{1}{\hbar^2 q^2/2m + \epsilon} \frac{dq}{2\pi} = 1. \qquad (6.25)$$

The largest contribution to the integral comes from the lower limit, and the exact value of the "ultraviolet" cutoff Λ does not matter. Thus, our previous approximation for $U_{\mathbf{k}}$ is self-consistent, and the details of the potential near $r = 0$ (at large k) do not affect the energy of a weakly bound state. We can extend the upper integration limit to infinity and obtain $\epsilon = mg^2/2\hbar^2$. Therefore, in one dimension, an attractive potential always has a bound state, and the energy of this bound state scales as

$$\epsilon \sim \frac{mR^2 V_0^2}{\hbar^2}. \qquad (6.26)$$

(b) In three dimensions, the integral turns into:

$$g \int_0^\Lambda \frac{1}{\hbar^2 q^2/2m + \epsilon} \frac{4\pi q^2 dq}{(2\pi)^3} = 1, \tag{6.27}$$

and the dominant contribution now comes from the upper limit, i.e., large momenta q. In the limit $\epsilon \ll \hbar^2 \Lambda^2/2m$ the integral is independent of ϵ, and the left-hand side is equal to $gm\Lambda/\pi^2\hbar^2$. Clearly there exists a small enough g for which the equation (6.27) can not be satisfied, which means that there is no bound state in that case. Dropping numerical factors of order unity, and substituting $g \approx V_0 R^3$, $\Lambda \approx 1/R$, we obtain the approximate condition for the existence of a bound state:

$$V_0 \gtrsim \frac{\hbar^2}{mR^2}. \tag{6.28}$$

The numerical pre-factor here depends on the exact shape of the potential $U(r)$. Note that this is exactly the estimate given in Eq. (6.15).

(c) In two dimensions, the integral turns into:

$$g \int_0^\Lambda \frac{1}{\hbar^2 q^2/2m + \epsilon} \frac{2\pi q dq}{(2\pi)^2} = 1, \tag{6.29}$$

which scales logarithmically with the value of the upper limit:

$$\frac{gm}{2\pi\hbar^2} \ln\left(\frac{\hbar^2\Lambda^2}{2m\epsilon}\right) = 1, \tag{6.30}$$

which gives

$$\epsilon \approx \frac{\hbar^2}{2mR^2} e^{-2\hbar^2/mR^2 V_0}. \tag{6.31}$$

The solution for energy ϵ exists for an arbitrarily small g. Our solution has "*logarithmic accuracy*," which means that the value of the exponent does not depend on Λ, nor on the exact shape of the potential near $r - 0$, while the numerical pre-factor does.

An alternative solution

(a) In order to make the symmetry of the problem more explicit, let us redefine the coordinate, so that the potential well is from $-R/2$ to $R/2$. Then the solution of our symmetric problem should be either a symmetric or an antisymmetric function. The wave function of the lowest bound state, however, has to be symmetric on the account of the *oscillation theorem* (see, for example, Landau and Lifshitz 1991) that says that the number of zero crossings at finite r should be equal to $n-1$, where n is the number of the state starting from the lowest ($n = 1$ in this case). With this, we immediately write the solution of the Schrödinger equation (with some arbitrary normalization):

$$\psi = A\exp(\kappa r), \; r < -R/2; \tag{6.32}$$
$$\psi = \cos(kr), \; -R/2 \le r \le R/2; \tag{6.33}$$
$$\psi = A\exp(-\kappa r), \; r > R/2, \tag{6.34}$$

where

$$\kappa = \frac{\sqrt{-2mE}}{\hbar}; \; k = \frac{\sqrt{2m(E + V_0)}}{\hbar}, \tag{6.35}$$

and $E < 0$ is the energy of the bound state.

The wave function and its derivative should both be continuous at $x = R/2$:

$$\cos(kR/2) = A\exp(-\kappa R/2), \tag{6.36}$$
$$k\sin(kR/2) = A\kappa \exp(-\kappa R/2), \tag{6.37}$$

which yields

$$\tan\left(\frac{kR}{2}\right) = \frac{\kappa}{k} = \sqrt{\frac{-E}{E + V_0}}. \tag{6.38}$$

This equation appears to always have a solution. Assuming a "shallow" level with energy such that $|E| \ll V_0$, we have $\tan(kR/2 \to 0)$, and correspondingly, $k \to 0$, which corresponds to the wave function extending far beyond the extent of the potential.

Before moving to the three-dimensional case, let us discuss the next lowest state. By the oscillation theorem, there the wave function should have one zero crossing, which is, by symmetry, at $r = 0$, so the solution is an antisymmetric function:

$$\psi = -A\exp(\kappa r), \; r < -R/2; \tag{6.39}$$
$$\psi = \sin(kr), \; -R/2 \le r \le R/2; \tag{6.40}$$
$$\psi = A\exp(-\kappa r), \; r > R/2. \tag{6.41}$$

Matching the boundary conditions as previously illustrated leads to

$$\cot\left(\frac{kR}{2}\right) = -\frac{\kappa}{k} = -\sqrt{\frac{-E}{E + V_0}}. \tag{6.42}$$

Assuming as before that the quantity on the left-hand side is small leads to $kR/2 \approx \pi/2$, which is nothing else but our condition (6.15). By adjusting the parameters of the potential, we can violate the condition and "expel" the state. The bottom line is that in a one-dimensional well, one state exists always, but this is not the case for higher lying states.

(b) As it turns out, our brief inquest into the existence of the first excited state in a one-dimensional well has also answered the question of whether there is always a bound state in a three-dimensional potential. First of all, we again assume that the wave function of lowest-energy bound state is a function of radius only. In other words, we assume that the angular momentum is zero as nonzero angular momentum only raises the energy of the particle. With this, the Schrödinger equation reads:

$$-\frac{\hbar^2}{2M}\frac{1}{r^2}\frac{\partial}{\partial r}\left(r^2\frac{\partial\psi}{\partial r}\right) + (V - E)\psi = 0. \tag{6.43}$$

Here the first term representing the kinetic energy contains the radial part of the Laplacian in spherical coordinates.

While this is essentially different from the one-dimensional Schrödinger equation, we can recover the latter by making the standard substitution

$$\psi(r) = \frac{u(r)}{r}, \tag{6.44}$$

which yields an equation for u:

$$-\frac{\hbar^2}{2M}\frac{\partial^2 u}{\partial r^2} + (V - E)u = 0. \tag{6.45}$$

An important boundary condition for u is that, if we wish $\psi(0)$ to be finite, then, according to Eq. (6.44), it must be that $u(0) = 0$. Thus, in fact, we have exactly the situation we discussed previously for the antisymmetric state. Of course, there is no such thing as $r < 0$ in three dimensions, but the antisymmetric solution of the Schrödinger equation is the same for $r > 0$. Thus, a bound state does *not* always exist in a three-dimensional well.

(c) The two-dimensional case is the trickiest of the three, and is rarely discussed in undergraduate quantum mechanics courses.

Assuming, we are dealing with a wave function $\psi(r)$ that does not have any angular dependence, the Schrödinger equation in the two-dimensional case can be written as

$$\frac{\partial^2\psi}{\partial r^2} + \frac{1}{r}\frac{\partial\psi}{\partial r} + \frac{2M(E - V)}{\hbar^2}\psi = 0. \tag{6.46}$$

The solution for the two regions in term of *special functions* is essentially given in the formulation of the problem:

$$\psi = J_0(kr), \ 0 \le r \le R; \tag{6.47}$$

$$\psi = C_3 H_0^{(1)}(i\kappa r), \ r > R. \tag{6.48}$$

Here we again use an arbitrary normalization by setting $C_1 = 0$, and set $C_2 = 0$ to avoid the divergence of the wave function at the origin. Matching the wave functions and their derivatives at $r = R$ assuming $kR \ll 1$ and $\kappa R \ll 1$ for a weakly localized state, we have:

$$1 \approx C_3 ln(\kappa R), \tag{6.49}$$

$$-\frac{k^2 R}{2} \approx -\frac{C_3}{R}. \tag{6.50}$$

Solving for C_3 in Eq. (6.50) and substituting into Eq. (6.49), we find

$$-\frac{k^2 R^2}{2} \ln(\kappa R) \approx 1, \tag{6.51}$$

from which, substituting k and κ from Eq. (6.35), we get

$$E \approx -\frac{\hbar^2}{2mR^2} \exp\left(-\frac{2\hbar^2}{mV_0 R^2}\right). \tag{6.52}$$

The pre-factor in this expression is the localization energy in a region of radius R, and the exponential contains the ratio of this energy to the depth of the potential well, a large number for shallow wells.

The bottom line is that there is always a bound state in an arbitrary shallow two-dimensional well, but the (negative) energy of this state is exponentially small.

A more detailed discussion of the existence of bound states in potential wells in two (and other) dimensions can be found in books by Landau and Lifshitz (1991), Galitski et al. (2013), and Band and Avishai (2012).

6.4   Hypothetical anomalous hydrogen

Consider an atom that has a proton and an electron just like the normal hydrogen, but with the interaction potential between the proton and the electron given by

$$U(r) = -\frac{C}{r^S},\tag{6.53}$$

where $C > 0$ is a constant with appropriate dimensions, and S is a dimensionless positive number.

(a) For what values of S is there a solution that qualitatively resembles the usual hydrogen, i.e., we have a finite-sized bound atom?

(b) Without worrying about numerical factors of order unity, figure out the analog of the first Bohr radius for this hypothetical atom. The answer should match the usual-hydrogen case for $S = 1$.

(c) What is the characteristic speed of the electron in our exotic atom?

(d) Suppose $S = 1 + \delta$, $|\delta| << 1$. What would be a manifestation of the anomaly? Can you propose an experiment to look for such an anomaly?

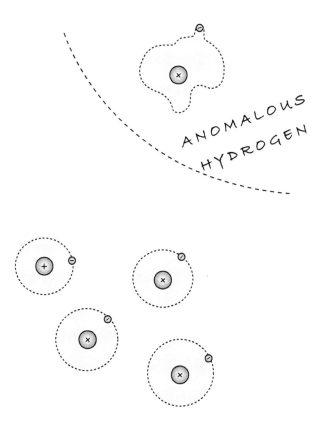

Solution

(a,b) In order to solve the problem, let us follow the reasoning that is frequently applied to the case of the usual hydrogen.

The potential energy of Eq. (6.53) decreases as the electron approaches the nucleus. However, when the electron is localized around the nucleus with a characteristic radius r, there is kinetic energy associated with such localization. Indeed, the *position-momentum uncertainty relation* tells us that a characteristic magnitude of momentum associated with the localization is

$$p = \frac{\hbar}{r}, \tag{6.54}$$

so that the corresponding kinetic energy is (we are neglecting numerical factors!)

$$K = \frac{p^2}{m} = \frac{\hbar^2}{mr^2}, \tag{6.55}$$

where m is the electron mass.

We see that shrinking the size of the atom r lowers the atom's potential energy, but increases the kinetic energy. The size of the ground state can be found by minimizing the total energy

$$E_{tot} = U + K = -\frac{C}{r^S} + \frac{\hbar^2}{mr^2} \tag{6.56}$$

with respect to r. (Since we are not careful about the numerical factors, there might be such factors neglected in front of each of the two terms in Eq. (6.56). This has no bearing on our present consideration, however.)

Let us examine the right-hand side of Eq. (6.56) in the limit $r \to 0$. If $S < 2$, the second (repulsion) term dominates in this limit, and the minimum of energy occurs at a finite value of r. Setting the derivative of the right-hand side of Eq. (6.56) with respect to r to zero, we have (again, neglecting the numerical factors)

$$\frac{SC}{r^{S+1}} - \frac{\hbar^2}{mr^3} = 0. \tag{6.57}$$

It follows that, for the characteristic radius of the atom (for which the total energy is minimum),

$$r^{2-S} = \frac{\hbar^2}{mSC}, \tag{6.58}$$

or

$$\boxed{r = \left(\frac{\hbar^2}{mSC}\right)^{\frac{1}{2-S}}.} \tag{6.59}$$

This answers part (b) of the problem. For the usual hydrogen, we have $S = 1$, $C = e^2$, where e is the magnitude of the electron charge. Substituting this into Eq. (6.59), reproduces the Bohr radius a_0:

$$r = a_0 = \frac{\hbar^2}{me^2}. \tag{6.60}$$

If $S > 2$, the first (attraction) term dominates in the limit of $r \to 0$, and the energy decreases without bound as r gets smaller. What this means is that the electron "falls" onto the nucleus, and we do not, in fact, have a finite-size atom. In the boundary case of $S = 2$, what happens depends on the value of C relative to $\hbar^2/m$. A detailed discussion will not be given here, but can be found in Sec. 35 of the book by Landau and Lifshitz (1991).

(c) The characteristic speed v of the electron can be found by substituting Eq. (6.59) into Eq. (6.54). Assuming the electron is nonrelativistic, we have

$$v = \frac{p}{m} = \frac{\hbar}{m}\left(\frac{\hbar^2}{mSC}\right)^{\frac{1}{S-2}} = \left(\frac{\hbar^S}{m^{S-1}SC}\right)^{\frac{1}{S-2}}. \tag{6.61}$$

For the usual hydrogen ($S = 1$, $C = e^2$), we recover

$$v = \frac{e^2}{\hbar} = \alpha c, \tag{6.62}$$

where c is the speed of light, and $\alpha = e^2/(\hbar c)$ is the *fine-structure constant*.

(d) In order to analyze the effect of the anomaly, let us first rewrite the constant C so that its dimensions are immediately transparent: $C = e^2 R^\delta$, where R is a constant of dimensions of length. Now we expand the potential energy of Eq. (6.53) in the small parameter δ up to first term in δ in the Taylor series (note, we keep track of all numerical coefficients in this part).

$$\begin{aligned} U = -\frac{C}{r^{1+\delta}} &= -\frac{e^2}{r}\left(\frac{R}{r}\right)^\delta \\ &\approx -\frac{e^2}{r}\left[1 + \delta \ln\left(\frac{R}{r}\right)\right] \\ &= -\frac{e^2}{r} + \delta\frac{e^2}{r}\ln\left(\frac{r}{R}\right) = -\frac{e^2}{r} + \hat{V}, \end{aligned} \tag{6.63}$$

where

$$\hat{V} = \delta\frac{e^2}{r}\ln\left(\frac{r}{R}\right) \tag{6.64}$$

is the perturbation term in the Hamiltonian that can be used to analyze the effect of the anomalous interaction on the hydrogen spectra. Comparing the predicted shifts of the energy levels with experimental data, we will be able to rule out certain ranges of the parameters δ and R characterizing the anomalous interaction.[1]

---

[1] The expansion (6.63) is valid only for a range of the values of r, for which $|\delta \ln(R/r)| \ll 1$. It would need to be verified that the contribution to the energy shifts from the very small and very large distances where this equality does not hold are negligible. Another thing to keep in mind is that the nucleus has finite size; therefore, there are deviations from the usual Coulomb potential at small distances, even in the absence of the hypothetical effect that we are discussing here.

We now apply first-order perturbation theory and calculate the matrix elements of the perturbation (6.64) for the $2S$ and $2P$ states in hydrogen. Since the perturbation does not depend on any angular variables, it is sufficient to consider only the radial integrals:

$$\Delta E_{2S} = \int_0^\infty \left[\frac{1}{\sqrt{2}} a_0^{-3/2} e^{-\frac{r}{2a_0}} \left(1 - \frac{r}{2a_0}\right)\right]^2 \delta \frac{e^2}{r} \ln\left(\frac{r}{R}\right) r^2 dr, \tag{6.65}$$

$$\Delta E_{2P} = \int_0^\infty \left[\frac{1}{2\sqrt{6}} a_0^{-3/2} e^{-\frac{r}{2a_0}} \frac{r}{a_0}\right]^2 \delta \frac{e^2}{r} \ln\left(\frac{r}{R}\right) r^2 dr. \tag{6.66}$$

Here ΔE_{2S} and ΔE_{2P} are the energy shifts due to the perturbation, and we have used the explicit form of the radial wave functions for hydrogen.

The integrals in Eqs. (6.65) and (6.66) can be evaluated analytically (for example, using *Mathematica*), with the following results (in atomic units, $me^4/\hbar^2$):

$$\Delta E_{2S} = \frac{\delta}{8} \left[3 - 2\gamma - 2\ln\left(R/a_0\right)\right], \tag{6.67}$$

$$\Delta E_{2P} = \frac{\delta}{96} \left[11 - 6\gamma - 6\ln\left(R/a_0\right)\right], \tag{6.68}$$

where $\gamma \approx 0.577$ is the *Euler–Mascheroni Constant*.

We see that, since the energy shifts (6.67) and (6.68) are different, the perturbation lifts the *accidental degeneracy* between the states of the same principal quantum number but different angular momentum, which is peculiar to the pure Coulomb potential. Since the degeneracy is already lifted by *fine-structure interactions* and *Lamb shift*, our anomalous interaction would lead to additional corrections to the energy intervals.

6.5   Time-reversal in quantum mechanics

Consider a quantum mechanical *transition matrix element*

$$\langle \psi_f | \hat{O} | \psi_i \rangle, \tag{6.69}$$

where ψ_i and ψ_f are the initial and final wavefunctions, and $\hat{O}$ is the *transition operator* sometimes also called *transition Hamiltonian*. For example, the operator for *electric-dipole (E1) transitions* is

$$\hat{O} = -\mathbf{d} \cdot \mathbf{E}, \tag{6.70}$$

where $\mathbf{d}$ is the dipole operator, and $\mathbf{E}$ is the electric field.

Discuss the properties of transition matrix element with respect to the operation of *time reversal*. Specifically, what happens to the real and imaginary parts of the transition matrix element upon time reversal?

Solution

The meaning of time reversal is that the initial state of the process becomes the final state and vice versa, so that the original matrix element becomes

$$\langle \psi_i | \hat{O} | \psi_f \rangle. \tag{6.71}$$

Assuming that $\hat{O}$ is a *Hermitian operator*, the result of the operator acting "on the left" is complex conjugate to the result of acting "on the right" since a *bra wavefunction* is complex conjugate of a *ket wavefunction*:

$$\langle \psi_i | \hat{O} = \left(\hat{O} | \psi_i \rangle \right)^*. \tag{6.72}$$

A consequence of this is that the transition matrix element upon time reversal is complex conjugate to the original matrix element. Therefore, the real part of a transition matrix element is invariant with respect to time reversal (is T *even*), while the imaginary part changes sign upon time reversal (is T *odd*).

6.6   Superconductivity vs. atomic diamagnetism

An atom contains electrons that rearrange, without resistance, when external fields are applied, and stationary atomic states may have electric currents associated with them that do not decay, which are thus *persistent currents*. It is therefore tempting to draw an analogy between atoms and superconductors. In this problem, we explore how well such analogy holds.

(**a**) Consider a superconducting sphere of radius a placed in a uniform external magnetic field $\mathbf{B}_0$. Find the induced magnetic moment of the sphere.

(**b**) Based on part (a) and your knowledge of the "facts of life" about atoms, can a diamagnetic atom (i.e., an atom with zero total angular momentum) be approximated as a superconducting sphere when it comes to the atom's diamagnetic response?

(**c**) Estimate, based on parts (a,b) and using *dimensional analysis*, the magnitude of the magnetic moment induced in a few-electron diamagnetic atom (e.g., helium, beryllium,...).

(**d**) Discuss a physical picture of how the induced moment of a diamagnetic atom comes about.

Hint: for the last part, recall the *Larmor theorem*, which states that for a system of charged particles, all having the same ratio of charge to mass, moving in a *central field*, the motion in a uniform magnetic induction $\mathbf{B}$ is, to first order in $\mathbf{B}$, the same as in the absence of $\mathbf{B}$ except for the superposition of a common precession around the direction of $\mathbf{B}$ with the Larmor frequency.

Solution

(a) Magnetic induction inside a superconductor is zero; in an external field, currents are induced on the surface that compensate the external field. The boundary condition for magnetic induction **B** is that its normal component is continuous through a boundary. Therefore, the induced surface currents are such that the normal component of the total field outside the superconductor is zero; the magnetic-field lines flow around the superconductor (Fig. 6.3).

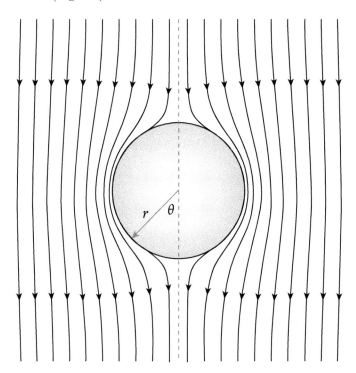

Fig. 6.3 Magnetic field lines expelled by a superconducting sphere. The magnetic induction inside the sphere is zero, and, for a uniform external field, the field outside is the superposition of the external field and a dipole field due to the currents flowing on the surface of the sphere.

For the present case of spherical geometry, we can (almost) guess the solution (cf. also Prob. 2.1). The magnetic induction due to the surface currents inside the sphere is uniform (and equal to $-\mathbf{B}_0$), and the field outside the sphere is that of a magnetic dipole $\boldsymbol{\mu}$:

$$\mathbf{B}_d = \frac{-\boldsymbol{\mu} + 3(\boldsymbol{\mu} \cdot \hat{r})\hat{r}}{r^3}, \tag{6.73}$$

where **r** is the radius vector from the center of the sphere. The boundary condition for the normal component of magnetic induction for a point near the surface outside the sphere with radius vector making an angle θ with the external field (Fig. 6.3) reads:

$$B_0 \cos\theta - \frac{-\mu\cos\theta + 3\mu\cos\theta}{a^3} = 0, \tag{6.74}$$

from which it follows that the induced magnetic moment of the sphere is:

$$\boxed{\boldsymbol{\mu} = -\frac{a^3 \mathbf{B}_0}{2}.} \tag{6.75}$$

A superconductor is an *ideal diamagnetic*, and the magnitude of the *diamagnetic polarizability* of a spherical superconductor is one half of the cube of its radius.

(b) The fact of life is that magnetic field penetrates inside atoms and molecules, which is a central fact in the field of *nuclear magnetic resonance (NMR)*. In fact, the electrons in diamagnetic atomic and molecular substances do affect the magnetic fields at the nucleus, but these *chemical shifts* are typically in the 1–100 *ppm* range, where ppm stands for parts per million. Thus, an atom is not a superconductor!

(c) Had the atom been a superconducting sphere, the estimate for the induced magnetic moment would have been $-a_0^3 B_0$, but this cannot be the correct answer based on the argument of part (b). It does have the right dimensions, so we must be off by some dimensionless factor much smaller than unity.

Let us try the following approach. The induced magnetic moment is proportional to B_0, and the energy of the atom due to the interaction of the induced dipole moment with B_0 is $\propto B_0^2/2$, where the factor of one half (which is irrelevant to our crude estimate, but is good to keep in mind, albeit it will be dropped in the further discussion) comes from integrating the external magnetic induction from zero to B_0. To end up with a quantity of the desired dimensions of energy, we multiply B_0^2 by the square of the characteristic atomic magnetic moment μ_B^2 (the square of *Bohr magneton*; $\mu_B = e\hbar/(2mc)$, where e and m are the magnitude of the electron's charge and its mass, respectively, and c is the speed of light), and divide by the characteristic atomic energy, the rydberg $me^4/(2\hbar^2)$. Combining all of this together, we have for the energy, ignoring all numerical factors:

$$\left(\frac{e\hbar}{mc}B_0\right)^2 \frac{\hbar^2}{me^4} = \frac{\hbar^4}{m^3 e^2 c^2} B_0^2 = \frac{e^4}{\hbar^2 c^2}\left(\frac{\hbar^2}{me^2}\right)^3 B_0^2 = \alpha^2 a_0^3 B_0^2, \tag{6.76}$$

where we have taken into account that the Bohr radius is $a_0 = \hbar^2/(me^2)$ and the fine-structure constant is $\alpha = e^2/(\hbar c)$. From this, we estimate the induced diamagnetic moment as

$$\boxed{\boldsymbol{\mu} = -\alpha^2 a_0^3 \mathbf{B}_0.} \tag{6.77}$$

Our dimensions-based previous estimate may give an impression that the energy shift of a diamagnetic atom in an external field is a *second-order perturbation-theory* effect. This also matches the intuitive picture where the magnetic field induces an atomic magnetic moment, and this *induced magnetic moment* interacts with the field producing the energy shift. Interestingly, in quantum mechanics, the effect comes not in the second order but in the *first order of perturbation theory* due to the term in the Hamiltonian for an atomic electron

$$\hat{H} = \frac{1}{2m}\left(\mathbf{p} + \frac{e}{c}\mathbf{A}\right)^2 + \hat{V} \tag{6.78}$$

(where $\mathbf{p}$ is the *momentum*, and $\hat{V}$ is the *potential-energy* term) that is quadratic in the vector potential $\mathbf{A}$ (Kittel, 2007). The diamagnetic energy shift is the expectation value of

$$\frac{e^2}{2mc^2}A^2. \tag{6.79}$$

(**d**) According to Larmor's theorem, an atom in a weak magnetic field is no different from an atom in the absence of the field, which means, in particular, that there is no mixing of any of the higher-lying atomic states to the ground state. The only effect of the magnetic field is to put the electrons into rotation with the Larmor frequency Ω_L. (In the frame co-rotating with the Larmor frequency, there is no first-order effect of the magnetic field at all.) The magnitude of current associated with the rotation can be estimated as

$$I \approx e\frac{\Omega_L}{2\pi} \approx \frac{e^2 \mathbf{B}_0}{2\pi mc}. \tag{6.80}$$

The magnetic moment associated with a loop current is proportional to the current and the loop area A:

$$\mu = \frac{IA}{c} \approx \frac{e^2 \mathbf{B}_0}{2\pi mc^2}\pi a_0^2 \approx \alpha^2 a_0^3 \mathbf{B}_0, \tag{6.81}$$

in agreement with the result of Eq. (6.77). It is also interesting to compare the result of our crude estimate with that of a more accurate calculation (see, for example, Kittel 2007):

$$\boldsymbol{\mu} = -\frac{Z}{6}\alpha^2 a_0 \langle R^2 \rangle \mathbf{B}_0, \tag{6.82}$$

where Z is the *atomic number* and $\langle R^2 \rangle$ is the mean square distance of the electrons from the nucleus.

6.7 Atomic desorption

(a) Consider atoms adsorbed on a surface that occasionally desorb from the surface due to *thermal activation*. What do you expect the angular distribution of the desorbed atoms to be? Formulate the answer in terms of relative probability of a desorbed atom moving at an angle θ with respect to the normal to the surface. Assume that the collisions in the gas phase can be neglected. Give a physical picture explaining your reasoning.

(b) Assuming that the probability of emission goes as $\cos\theta$ (the *Knudsen Law*), what is the average distance that an atom contained in a spherical cell of radius R travels between wall collisions? Assume that the atom is adsorbed upon each wall collision and then subsequently desorbed.

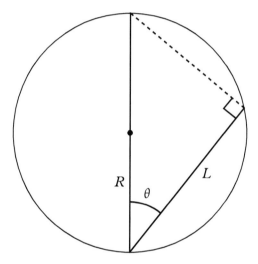

Fig. 6.4 Atomic *desorption* from the inner wall of a spherical vapor cell.

Solution

(a) Imagine that an atom is adsorbed on a surface patch with an area dA. To determine the probability of the atom flying off in a certain direction forming the angle θ with the normal to the surface, we can reasonably assume that this probability scales as the effective area of the surface element, i.e., $dA \cos \theta$, as seen from the direction of the emission. This is similar to the inverse: the probability of a gas atom impinging on a surface hitting a surface element scales as the cosine of the angle between the atom's direction and the normal to the surface. Experiments show that the cosine distribution is usually an excellent approximation.

(b) Assuming the $\cos \theta$ distribution (the Knudsen Law), the probability of atomic desorption at an angle θ is proportional to

$$dP(\theta) \propto \cos \theta 2\pi \sin \theta d\theta, \qquad (6.83)$$

where we took into account the solid-angle factor and the independence of the emission probability on the azimuth.

If an atom is emitted at an angle θ (Fig. 6.4), it will travel the distance $L = 2R \cos \theta$ till it hits the wall of the spherical cell once again. To find the average distance between wall collisions, we integrate the product of the emission probability for different values of θ and the corresponding travel distance:

$$\bar{L} = \frac{\int_0^{\pi/2} (2R \cos \theta) \cos \theta 2\pi \sin \theta d\theta}{\int_0^{\pi/2} \cos \theta 2\pi \sin \theta d\theta}, \qquad (6.84)$$

where, in the denominator, we have the same expression as in the numerator but without the distance, which is needed to properly normalize the probability.

Evaluating the elementary integrals, we find for the average distance:

$$\boxed{\bar{L} = \frac{4R}{3}.} \qquad (6.85)$$

As it turns out, a general result can be formulated for a cell of arbitrary shape (not just a sphere) with inner volume V and inner surface area S:

$$\boxed{\bar{L} = \frac{4V}{S}.} \qquad (6.86)$$

of which the result of Eq. (6.85) is a particular case.

Can you derive Eq. (6.86)?

6.8   Lamb shift

In this problem we estimate the *Lamb shift* of the 2s state of the hydrogen atom.[2]

(a) The Lamb shift is due to zero-point *vacuum fluctuations* of the electromagnetic field. Assume the atom is inside a box of volume V. What is the mean-squared amplitude $\overline{E^2}_\omega$ of the electric field of these fluctuations for an electromagnetic field mode at frequency ω?

The fluctuating electric field calculated in part (a) perturbs the motion of the electron in the hydrogen atom. One can estimate the mean squared electron displacement $\overline{r^2}_\omega$ under the influence of $\overline{E^2}_\omega$, and sum over all electromagnetic field modes to find the total displacement $\overline{r^2}$. As a result of this displacement, there is a perturbation to the Coulomb potential acting on the electron, the resulting energy level shifts are calculated in first-order perturbation theory.

(b) Write down the equation of motion for the electron under the influence of the electric field $\overline{E^2}_\omega$, and calculate the resulting displacement $\overline{r^2}_\omega$ of its trajectory.

(c) Sum over all electromagnetic field modes to estimate the total r.m.s. perturbation $\overline{r^2}$ of the trajectory of the electron.

(d) The electron moves in the Coulomb potential of the nucleus $V(r) = -e^2/r$. What is the perturbation to this potential due to fluctuations $\overline{r^2}$ of the trajectory of the electron?

(e) Use first-order perturbation theory to find the shift of the 2s energy level due to this perturbation.

---

[2]This problem was inspired by the Lamb-shift treatment by Migdal and Krainovl (1969).

Solution

(a) The energy of zero-point vacuum fluctuations of the electromagnetic field for a mode with angular frequency ω is given by

$$U_\omega - \frac{1}{2}\hbar\omega. \tag{6.87}$$

The average values of fluctuating electric and magnetic fields vanish, however they have nonzero mean-squared values $\overline{E^2} = \overline{B^2} \neq 0$. For each field mode the energy density is

$$\frac{U_\omega}{V} = \frac{\overline{E^2}_\omega + \overline{B^2}_\omega}{8\pi} = \frac{\overline{E^2}_\omega}{4\pi}. \tag{6.88}$$

Therefore, the mean-squared fluctuating electric field is

$$\overline{E^2}_\omega = \frac{2\pi}{V}\hbar\omega. \tag{6.89}$$

(b) These fluctuating electromagnetic fields lead to a jitter of the electron position, historically called with a German word *Zitterbewegung*. Let us consider the motion of an electron in hydrogen atom, acted upon by this fluctuating electric field. At frequencies much lower than atomic frequency $\hbar\omega_a = 2\,Ry = \alpha^2 mc^2$, the electron's motion is determined by the Coulomb potential, and the jitter is small. To find this jitter at frequencies much higher than atomic frequencies, we can neglect the atomic Coulomb potential and consider the electron as free. The equation of motion for electron position **r** is:

$$m\ddot{\mathbf{r}} = e\mathbf{E}. \tag{6.90}$$

The electron displacement due to electric field in mode ω is then given by $m\omega^2 r_\omega = eE_\omega$. The mean $\overline{r_\omega}$ vanishes, and the mean-squared displacement is

$$\overline{r^2}_\omega = \frac{e^2}{m^2\omega^4}\overline{E^2}_\omega = \frac{2\pi}{V}\frac{\hbar e^2}{m^2\omega^3}. \tag{6.91}$$

(c) There is no interference between different electromagnetic field modes (since their corresponding fluctuating displacements are at different frequencies), thus, to find the total displacement squared $\overline{r^2}$ we have to sum the mean-squared displacements due to all modes:

$$\overline{r^2} = \int \overline{r^2}_\omega \rho(\omega) d\omega, \tag{6.92}$$

where the mode density $\rho(\omega)$ is given by

$$\rho(\omega)d\omega = 2V\frac{4\pi k^2 dk}{(2\pi)^3} = V\frac{\omega^2 d\omega}{\pi^2 c^3}, \tag{6.93}$$

where the factor of 2 appears when summing over the two independent photon polarizations. Therefore,

$$\overline{r^2} = \frac{2\hbar e^2}{\pi m^2 c^3} \int \frac{d\omega}{\omega}. \tag{6.94}$$

We note that this integral diverges both at the lower and at the upper limits—such infinities are a common occurrence in *quantum electrodynamics*. However, there are physically motivated frequency cutoffs.

As the upper limit we use the electron Compton frequency ω_C, given by $\hbar\omega_C = mc^2$. This can be motivated by the picture of the interaction with the electromagnetic field as a sequence of emission and absorption of the virtual photons with energy $\hbar\omega$. It is clear that our nonrelativistic treatment of the electron motion fails in this case. At any rate, whatever happens at such high frequencies should be the same for the bound and unbound electrons, and is an issue of the "renormalization" of the electron mass rather than the estimate of the Lamb shift.

As the lower limit we use the atomic frequency ω_a, defined previously. The motivation for this is somewhat subtle (though it is often brushed off in books discussing the estimate). Below this frequency the free-electron approximation (6.90) is no longer valid. We note that the estimates of the value of a logarithm are "tolerant" to errors in the argument.

We can now do the integral

$$\overline{r^2} = \frac{2\hbar e^2}{\pi m^2 c^3} \ln\left(\frac{1}{\alpha^2}\right). \tag{6.95}$$

(d) Let us recap: the vacuum fluctuations of the electromagnetic field give rise to "jitter" of the electron with vanishing average displacement $\overline{r} = 0$, but mean-squared displacement given by Eq. (6.95). In a hydrogen atom, the electron moves in the Coulomb potential $V(r) = -e^2/r$. By symmetry, $\overline{x^2} = \overline{y^2} = \overline{z^2} = \overline{r^2}/3$. Due to oscillations, the effective potential seen by the electron becomes

$$V(r) + \frac{1}{2}\left(\overline{x^2}\frac{d^2V}{dx^2} + \overline{y^2}\frac{d^2V}{dy^2} + \overline{z^2}\frac{d^2V}{dz^2}\right) = V(r) + \frac{1}{6}\overline{r^2}\nabla^2 V(r). \tag{6.96}$$

The second term gives the perturbation to the Coulomb potential of the hydrogen atom.

(e) Using first-order perturbation theory, the energy-level shift for a state $|\psi\rangle$ is $\Delta = \overline{r^2}\langle\psi|\nabla^2 V(r)|\psi\rangle/6$. Here, we can make use of the fact that $\nabla^2 V(r) = -4\pi e^2 \delta(\mathbf{r})$ to obtain

$$\Delta = -\frac{4\pi}{6}e^2|\psi(0)|^2\overline{r^2}. \tag{6.97}$$

We see that, within the adopted approximations, the energy shift vanishes for states with nonzero orbital angular momentum, which have $\psi(0) = 0$. For an s-state with principal quantum number n,

$$|\psi(0)|^2 = \frac{1}{\pi(na_0)^3} = \frac{1}{\pi}\left(\frac{\alpha mc^2}{n\hbar c}\right)^3, \tag{6.98}$$

where a_0 is the Bohr radius. Substituting into Eq. (6.97), we obtain the answer for the Lamb shift of an s-state with principal quantum number n:

$$\begin{aligned}
\Delta &= -\frac{4}{3\pi}\alpha^5\frac{mc^2}{n^3}\ln\left(\frac{1}{\alpha^2}\right) \\
&= -\frac{8}{3\pi}\alpha^3\frac{Ry}{n^3}\ln\left(\frac{1}{\alpha^2}\right). \tag{6.99}
\end{aligned}$$

For the 2s state of the hydrogen atom, our estimate gives $\Delta = 5.5 \times 10^{-6}$ eV $= 1.3$ GHz. This is within 20% of the observed shift of 1.057 GHz.

6.9 Van der Waals interaction

Consider two neutral hydrogen atoms, separated by distance R. Let us approximate them as two-level systems with energy splitting $\hbar\omega_0 \approx e^2/a_0$, where a_0 is the Bohr radius. When the interatomic separation R is much greater than the Bohr radius a_0 (so there is no appreciable overlap between their electrons), but less than c/ω_0, the two atoms experience *van der Waals interaction*.

(a) Is this interaction attractive or repulsive? How does the interaction potential scale with separation R?

(b) Estimate the magnitude of the van der Waals interaction potential between the atoms, without calculating any factors of order unity (such as angular integrals).

(c) When the atoms are far apart, $R \gg c/\omega_0$, the interaction is called the *Casimir–Polder interaction*, and its potential scales as $1/R^7$. Discuss why the physics is different in this regime. Using *dimensional analysis*, explain the $1/R^7$ scaling.

Solution

(a) The van der Waals interaction is attractive and scales as $1/R^6$.

(b) It is possible to evaluate the van der Waals interaction potential using nonrelativistic second-order perturbation theory. We outline this calculation first, and then discuss the physics of how this interaction comes about.

For the purposes of this estimate, let us neglect all the angular and numerical factors and work with magnitudes, not vectors. Suppose, at some instant in time, the nucleus and the electron of the first atom are separated by r_1 (one could say that r_1 is the position operator of the electron of the first atom). Then the electric field created at the position of the second atom is $E \approx er_1/R^3$. The interaction Hamiltonian of the second atom with this field is

$$H' = er_2 E = \frac{e^2}{R^3} r_1 r_2. \tag{6.100}$$

We use second-order perturbation theory to evaluate the change in the energy of the system of two atoms as a result of this interaction:

$$\Delta E = -\sum_{m,n} \frac{\langle 00| H' |mn\rangle \langle mn| H' |00\rangle}{E_{mn} - E_{00}}, \tag{6.101}$$

where the sum is over all the excited states of the system $|mn\rangle$. Since we approximate the atoms as two-level systems (levels $|0\rangle$ and $|1\rangle$), the excited states are: $|10\rangle$, $|01\rangle$, and $|11\rangle$. However, the only non-zero matrix elements are for the last state: $\langle 11| r_1 r_2 |00\rangle \approx a_0^2$ (a single-atom matrix element of $\mathbf{r}$ vanishes unless it connects two quantum states with an allowed electric-dipole transition, in this case the ground and the excited states). Thus, the van der Waals interaction energy between the two atoms is

$$\boxed{U(R) = \Delta E \approx -\frac{e^4 a_0^4}{\hbar \omega_0} \frac{1}{R^6} \approx -\frac{e^2}{a_0} \frac{a_0^6}{R^6},} \tag{6.102}$$

where we used $\hbar \omega_0 \approx e^2/a_0$.

Now let us discuss the physical picture of this interatomic interaction. The two atoms are neutral, so there is no Coulomb force between them. However, the atoms are polarizable, which can lead to a *dipole-dipole interaction* between them. The problem is that, in the ground state, the expectation value for the dipole moment of each atom vanishes, so why should this interaction be present?

This apparent paradox is resolved, in *quantum electrodynamics (QED)*, by the existence of vacuum *zero-point fluctuations* of the electromagnetic field. The vacuum state is the state of lowest energy, which means that the expectation values of the electric and magnetic fields vanish. However, the vacuum state is an eigenstate of the Hamiltonian, and not of the electric field ($\mathbf{E}$) and magnetic field ($\mathbf{B}$) operators, so the electric and magnetic fields (in each electromagnetic mode with frequency ω and mode volume V) fluctuate, around the mean values of zero, with $\langle \mathbf{E}_\omega^2 \rangle = \langle \mathbf{B}_\omega^2 \rangle = 2\pi \hbar \omega/V$, which results in the energy of $\hbar \omega/2$ in each electromagnetic mode. This fluctuating electric field is what polarizes the atoms, inducing the fluctuating dipole moments. The dipole moment of one atom creates an electric field that polarizes the other atom, and the dipole-dipole interaction between them is attractive.

A curious reader might wonder: the dipoles induced in both atoms by the low-frequency fluctuating electromagnetic fields of the zero-point fluctuations are fully correlated (i.e., are essentially the same). Where is then the $1/R^3$ term arising from the dipole-dipole interactions?

Let us look at the expression for the interaction energy of two identical dipoles $\mathbf{d}$ separated by $\mathbf{R}$:

$$U_3 = \frac{d^2\left(-1 + 3\cos^2\theta\right)}{R^3}, \tag{6.103}$$

where θ is the angle between $\mathbf{d}$ and $\mathbf{R}$. But the average value of $\cos^2\theta$ over all possible relative orientations of two vectors in three-dimensional space is precisely $1/3$, which ensures that $\langle U_3 \rangle$ vanishes when the atoms are in free space where vacuum fluctuations inducing the dipoles are the same in all directions.

A detailed treatment of this and other QED effects is given by Milonni (1994).

We note, however, that one can obtain the correct result for the van der Waals interaction, without referencing zero-point fluctuations, simply by applying second-order non-relativistic perturbation theory, as we have done previously.

(c) In part (b), we used non-relativistic perturbation theory, and assumed that the interaction between the two atoms is "instantaneous." However, at large separations ($R \gtrsim c/\omega_0$), it is important to take into account the finite speed of light, i.e., *retardation*. A correct way to do this, using QED, is to draw the Feynman diagrams for the interaction and calculate the corresponding matrix elements. Instead, here we make a dimensional argument about the scaling of the interaction in the limit $R \gg c/\omega_0$.

In part (a), the atomic dipoles are induced by *vacuum fluctuations* of the electric field, with a wide frequency spectrum. But the greatest response is at the atomic resonance frequency ω_0, where the fluctuating field is on resonance. When $R \ll c/\omega_0$, the two dipoles are phased with each other, and the previous, non-relativistic, treatment is sufficient. But, when $R \gg c/\omega_0$, the induced dipoles are correlated only at frequencies $\omega \lesssim c/R \ll \omega_0$, whereas the interaction due to higher-frequency dipole moment fluctuations is suppressed, since, by the time the electromagnetic field propagates from one atom to the other at the speed of light, the two dipoles are "out of phase." Therefore, only the low-frequency, "static" *electric polarizability* of each atom can affect the interaction. The static electric polarizability of each atom is $\approx a_0^3$, and, since we deal with the interaction of two induced dipoles, the polarizability must be squared. The Casimir–Polder force is a quantum and relativistic effect, thus, for the dimensional analysis, we have two fundamental constants to work with: $\hbar$, and c. In order to make a quantity with units of energy, we must combine $\hbar c/R$, and, adding the atomic polarizabilities, we get the scaling

$$\boxed{U(R) \approx -\frac{\hbar c}{R}\frac{a_0^6}{R^6} = -\hbar c\frac{a_0^6}{R^7}.} \tag{6.104}$$

6.10 Vacuum birefringence

A well-known effect in optics is *birefringence* induced by a magnetic field applied to a medium transversely to the direction of light propagation (Fig. 6.5a). Let us assume a medium that is isotropic in the absence of external fields. In order to detect the effect, one can use linearly polarized light and direct the magnetic field at $\pi/4$ to the light-polarization plane. The resultant polarization *ellipticity*, which is quadratic in the applied magnetic field, is detected at the output of the medium.

It turns out that *quantum electrodynamics (QED)* predicts a nonzero effect even in the absence of a medium, although the magnitude of the induced ellipticity is vanishingly small, on the order of $10^{-24} \cdot B^2/\mathrm{T}^2 \cdot l/\mathrm{m}$ rad. Here l is the length over which the magnetic field is applied.

(a) Sketch a possible *Feynman diagram* corresponding to the vacuum-birefringence process.

(b) Another *magneto-optical effect* is *Faraday rotation* of the plane of the linear polarization of light in a longitudinal magnetic field (Fig. 6.5b). The angle of Faraday rotation is usually linear in the applied field. Can there be Faraday effect in vacuum? Sketch a possible Feynman diagram and use it for your argument. Are there any fundamental symmetries that would be violated if such effect existed?

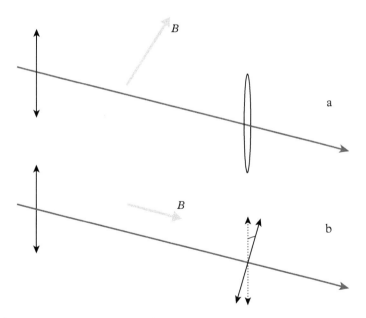

Fig. 6.5 (a) Magnetically induced birefringence. The output light ellipticity is quadratic in the applied transverse field. QED predicts that even vacuum can display such birefringence. (b) Faraday rotation. The angle or rotation of linear polarization is normally linear in applied longitudinal magnetic field.

Solution

(a) The change of light polarization can be thought of as a process where the incoming photon is absorbed, and a photon with the new polarization is emitted. But how can this happen in vacuum? A photon can decay into a *virtual electron-positron pair* (Fig. 6.6a), and a new photon can be born from the annihilation of this pair. The effect of the applied magnetic field is through electromagnetic interaction with the virtual electron and positron. This interaction is shown in Fig. 6.6a as virtual-photon lines corresponding to the static field. The fact that the field is static is schematically shown with virtual-photon lines terminating with an X.

(b) For Faraday rotation, we have a similar diagram (Fig. 6.6b) as that for vacuum birefringence, except there is only one photon from the external field on account of the fact that the effect should be linear in the magnetic field. It turns out that the transition amplitude corresponding to such a diagram is identically zero in QED. This follows from the fact that a photon is a particle with well-defined *C-parity*, where C stands for *charge conjugation*, and the fact that C-parity is conserved in electromagnetic interactions. Since the C-parity of a photon is -1, an even number of photons (e.g., the initial light photon and the one magnetic-field photon in the hypothetical vacuum Faraday effect) cannot transform into an odd number of photons (the single output photon). No such prohibition exists for the vacuum-birefringence process in Fig. 6.5a.

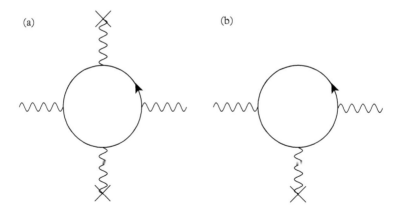

Fig. 6.6 (a) Feynman diagram for magnetically induced vacuum birefringence. On Feynman diagrams, time is going left to right. (b) Same for the hypothetical Faraday rotation. Such diagram corresponds to a zero amplitude in QED due to C-parity conservation for photons.

6.11   Nonmagnetic molecule

(a) *Electronic terms* of *diatomic molecules* are notated $^{2S+1}\Lambda_\Omega$, where S is the total electron spin, $\Lambda = 0, 1, 2, \ldots$ is the magnitude of the projection (in units of $\hbar$) of the total electron orbital angular momentum on the *molecular axis*, and Ω is the magnitude of the projection of the total electron angular momentum on this axis.

Explain why $^3\Delta_1$ molecules do not have a magnetic moment.

(b) Compare this with the case of $^3D_1$ atoms.[3]

[3]This problem was suggested to us by Amar Vutha.

Solution

(a) The magnitude of the projection of the total electronic orbital angular momentum on the molecular axis in a $^3\Delta_1$ state is 2, and that of the total electronic angular momentum is 1. This means that the projection Σ of the spin on the molecular axis [which, in contrast to the standard convention for Λ, can be either positive or negative (Herzberg, 1989)] must be $\Sigma = -1$.

The magnetic moment of the molecule in the molecular frame is along the molecular axis and its magnitude is given by $-\mu_B(\Lambda + 2\Sigma)$, under the approximation that the quantum numbers involved are "good."[4] The minus sign is because the *Bohr magneton* $\mu_B = e\hbar/(2mc)$ is defined with the magnitude of the electron charge, e. In the case of $^3\Delta_1$ states, the magnetic moment is zero. Note that the same is true for any state with $\Lambda = -2\Sigma$, for instance, $^2\Pi_{1/2}$.

(b) The magnetic moment for an atom can be calculated in a similar manner starting from

$$\boldsymbol{\mu} = -\mu_B(\mathbf{L} + 2\mathbf{S}) = -\mu_B(\mathbf{J} + \mathbf{S}), \tag{6.105}$$

where $\mathbf{L}, \mathbf{S}$, and $\mathbf{J}$ are the atom's orbital angular momentum, spin, and total angular momentum, respectively. The expectation value of the magnetic moment is along that of the total angular momentum $\mathbf{J}$, according to

$$\boldsymbol{\mu} = -g\mu_B\mathbf{J}, \tag{6.106}$$

where g is the *Landé g-factor*. To find g, according to Eq. (6.105), we need to find the expectation value of $\mathbf{S}$:

$$\mathbf{L} = \mathbf{J} - \mathbf{S} \implies \mathbf{L}^2 = (\mathbf{J} - \mathbf{S})^2 \implies \tag{6.107}$$

$$L(L+1) = J(J+1) + S(S+1) - 2\langle\mathbf{J}\cdot\mathbf{S}\rangle \implies \tag{6.108}$$

$$\langle\mathbf{J}\cdot\mathbf{S}\rangle = \frac{J(J+1) + S(S+1) - L(L+1)}{2}. \tag{6.109}$$

Taking the scalar product of both sides of Eq. (6.106) with $\mathbf{J}$ and taking the expectation values, we find

$$g = \frac{3}{2} + \frac{S(S+1) - L(L+1)}{2J(J+1)}. \tag{6.110}$$

For a $^3D_1$ state, $S = J = 1$ and $L = 2$, so that $g = 1/2$, which is relatively small, but certainly nonzero!

The difference between the atomic and molecular cases lies in the different underlying symmetry of the problem: central symmetry for atoms and axial symmetry for diatomic molecules. This leads to a difference in how we apply the *vector model* in the two cases: in molecules we just add projections of the angular momenta, while in atoms, we add full vectors, and then project them onto the direction of the expectation value of $\mathbf{J}$. The resulting average projections are generally different from the integer or half-integer projections we have in the molecular case. A tutorial discussion of the vector model can be found in Ch. 2 of the book by Auzinsh et al. (2010).

---

[4]Specifically, this refers to molecules where angular momenta couple according to *Hund's case* (a) (Herzberg, 1989).

6.12 Quantum mechanics of angular momentum

A quantum system with total angular momentum J is prepared in a state with a well-defined projection m_z of the angular momentum along **z**.

(a) What are the *expectation values* of the angular-momentum projection $\langle \hat{J}_x \rangle$ and its square $\langle \hat{J}_x^2 \rangle$?

(b) If one performs a single measurement of m_x, what, in general, are the possible outcomes of such a measurement?

(c) Let us specialize the discussion to the $J = 1$, $m_z = 0$ state. If we write this state in the basis where the quantization axis is **x**, what is the amplitude of the $m_x = 0$ sublevel?

Solution

(a) Already from the symmetry of the problem, it is clear that the average projection on **x** is zero, i.e., $\boxed{\langle \hat{J}_x \rangle = 0.}$ As for the expectation value of the square of the x-projection, this can be found by applying the *vector model* (Fig. 6.7). We write $J(J+1) = m_z^2 + J_\perp^2$, where $J_\perp^2 = J_x^2 + J_y^2$ is the square of component of **J** perpendicular to **z**. Again, from symmetry, $\langle J_\perp^2 \rangle = \langle J_x^2 \rangle + \langle J_y^2 \rangle = 2\langle J_x^2 \rangle$, from which we obtain

$$\boxed{\langle J_x^2 \rangle = \frac{J(J+1) - m_z^2}{2}.}$$
(6.111)

(b) In general, a possible outcome of the measurement is a projection m_x belonging to the set $-J, -J+1, \ldots, J$.

(c) Substituting $J = 1$, $m_z = 0$ into Eq. (6.111), we find that $\boxed{\langle J_x^2 \rangle = 1.}$ According to part (b), the possible outcomes for the measurement of m_x are $-1, 0, 1$. However, to get the expectation value of $\langle J_x^2 \rangle = 1$, it must be that one never measures $m_x = 0$. This means that the probability and the amplitude of this state are zero.

There are, of course, many different ways to solve the problem. One systematic way is with a *rotation matrix* as discussed in detail, for example, in the books by Budker et al. 2008 and Auzinsh et al. 2010.

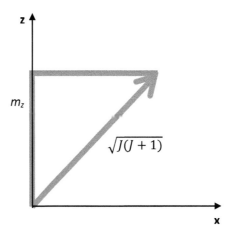

Fig. 6.7 A state of angular momentum J with a fixed projection m_z. The total length of the angular momentum vector is the square root of the expectation value of $\hat{\mathbf{J}}^2$.

6.13 Light shifts

The ground electronic states of *alkali atoms* have two *hyperfine states* (Fig. 6.8) with *total angular momentum* $F = I \pm 1/2$, where I is the *nuclear spin*. For example, for ^{87}Rb, $I = 3/2$ and $F = 2$ or 1.

Suppose that off-resonant circularly polarized light propagating along the quantization axis is applied to ^{87}Rb atoms. Due to the so-called *AC-Stark effect* a.k.a. *light shift*, the Zeeman sublevels of the hyperfine states shift by an amount that is proportional to the intensity of the light. The overall light shift (see the inset in Fig. 6.8) can be separated into *scalar light sift* independent of the hyperfine state and the *magnetic quantum number* M_F, and *vector light shift* which, for a given F, is proportional to M_F. In the absence of hyperfine structure, vector light shift is proportional to the projection of the *total electronic angular momentum* J, M_J.

What is the relation between the vector light shifts in the two hyperfine states assuming that the light detuning from resonance greatly exceeds the energy splitting between the hyperfine states?[5]

Hint: No explicit calculations should be necessary in this problem.

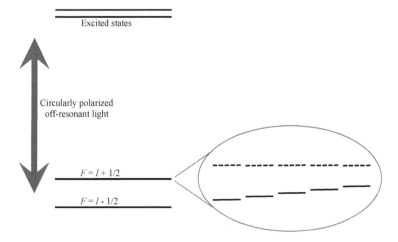

Fig. 6.8 Alkali atoms have two ground-state hyperfine levels with $F = I \pm 1/2$. When circularly polarized off-resonant light is applied, Zeeman sublevels shift as illustrated in the inset for an $F = 2$ state. The overall displacement of all sublevels is the scalar light shift while the splitting is due to the vector light shift.

[5]This problem was inspired by a discussion with Professor Ron Folman.

Solution

The answer to the question, namely, that the vector light shifts in the two hyperfine states are equal in magnitude and opposite in sign, becomes transparent upon two key realizations.

First, to a very good approximation, light does not directly affect nuclear states; its effect on the atom is due to the interaction with electrons. Thus, we can figure out how the light affects the eigenstates of the *total electronic angular momentum* M_J and then work out the shifts of the M_F states by expanding the $|F, M_F\rangle$ states in the $|J, M_J\rangle|I, M_I\rangle$ basis.

Second, even this we do not need to do explicitly as long as we realize that the situation with light shifts is exactly analogous to that with the *Zeeman effect*. Here too, the M_J states are predominantly shifted by the magnetic fields, while the *Landé g-factors* g_F reflect the composition of the corresponding states in terms of the $|J, M_J\rangle|I, M_I\rangle$ basis.

The result for the two hyperfine ground states for an alkali atom is that g_F are opposite for the two hyperfine states [see, for example, Prob. 2.4 in the book by Budker et al. (2008)].

We thus see that off-resonant circularly polarized light acts (apart from the scalar light shift) as an *effective magnetic field*, including its influence on the energies of the sublevels of the hyperfine states.

6.14 Optical pumping

Consider atoms with total angular momentum $F = 1$ in the ground state and $F' = 0$ in the excited state. If an atom is excited to the upper state, it decays back to the ground state with a 100% *branching ratio*. Initially, the population is uniformly distributed among the ground-state magnetic sublevels (1/3 of the population in each; Fig. 6.9) and there are no coherences. Now suppose that the atoms are excited with linearly polarized light that is turned on at some point in time. For concreteness, we assume that the light propagates along the quantization axis ($\hat{z}$) and the polarization is along $\hat{x}$. Assume that there is no *relaxation* among the ground-state sublevels and that there are no external fields other than the light. Due to *optical pumping* by the light, the population of atoms redistributes among the atomic sublevels.

What are the populations of the ground-state sublevels after a sufficiently long time when the populations have settled to their new steady-state values?

Fig. 6.9 An atom with total angular momentum $F = 1$ in the ground state with *Zeeman sublevels* with $M_F = -1, 0, 1$ and an excited state with $F' = 0$. Initially all ground-state sublevels are equally populated and there are no coherences. Figure courtesy of Simon M. Rochester, generated with the ADM package (http://rochesterscientific.com/ADM).

Solution

The answer is (1/4, 1/2, 1/4) for $M_F = -1, 0$, and 1, respectively, but a vast majority of scientists, even those working in laser spectroscopy and light-atom interactions, do not get this right.

In order to see what is going on, we first note that, with the chosen coordinate frame, the linearly polarized laser light corresponds to a coherent superposition of σ^+ and σ^- circularly polarized light driving the $M_F = -1 \rightarrow F' = 0$ and $M_F = 1 \rightarrow F' = 0$ transitions, respectively (Fig. 6.10, left). The light drives atoms from the $M_F = \pm 1$ sublevels into the excited state. Once atoms are excited, they decay with equal probability to each of the three ground-state sublevels.

Reaching this point in the analysis, one may (erroneously) conclude that the final population distribution is (0, 1, 0), so that all atoms accumulate in the $M_F = 0$ sublevel, which is a *dark state* in the sense that it does not interact with the light and so that the atoms accumulated there are not re-excited. What is missed here is that, in fact, there is not one but two dark states in this system!

In order to see this most clearly, let us look at the very same problem but changing the axes in such a way that the light polarization (rather than propagation direction) is along $\hat{z}$ (Fig. 6.10, middle). Here we see that only the $M_F = 0$ sublevel interacts with the light and the population gets equally distributed in the two *dark states*, in this case, the $M_F = -1$ and $M_F = 1$ sublevels.

We emphasize that no statement such as "there are two dark states in the system and the population gets equally distributed among them" should depend on the chosen basis, even though what these dark states actually are in terms of magnetic sublevels is, of course, basis dependent.

Going back to our original basis, the two dark states are 1) $F = 0$ and 2) a coherent superposition of $M_F = -1$ and $M_F = 1$ sublevels, which are, by symmetry, represented in this superposition with equal weight. (Of course, transforming the result between bases can be also done formally using *rotation matrices*, but this is not necessary to solve this problem.)

The final result is shown in Fig. 6.10, right.

The concept of *dark states* is essential for much of modern atomic physics, including such phenomena and techniques as *electromagnetically induced transparency (EIT), coherent population trapping, stimulated Raman adiabatic passage (STIRAP)* and many others.

A detailed discussion of the topics touched upon in this problem can be found, for example, in the books by Auzinsh et al. (2010) and Budker et al. (2008).

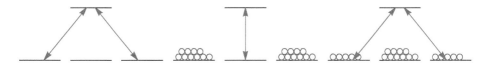

Fig. 6.10 Left: light that is linearly polarized orthogonally to $\hat{z}$ is a coherent super-position of circularly polarized components driving the $M_F = -1 \rightarrow M' = F' = 0$ and $M_F = 1 \rightarrow M' = F' = 0$ transitions. Middle: light polarized along $\hat{z}$ drives the $M_F = 0 \rightarrow M'_F = 0$ transition and depopulates the $M_F = 0$ sublevel. Right: the final population distribution of light polarization orthogonal to $\hat{z}$. The $M_F = \pm 1$ sublevels are populated because there is a coherent superposition of these states that is "dark" mean-ing that the atoms in such a state do not interact with the light. Figure courtesy of Simon M. Rochester, generated with the ADM package (`http://rochesterscientific.com/ADM`).

7
Nuclear and Elementary-Particle Physics

A neutron walks into a bar and asks
the bartender: "How much for a
drink?" Bartender looks at him and
says, "For you, no charge."

*Physics on Your Feet: Berkeley Graduate Exam Questions: or Ninety Minutes of Shame but a PhD for the Rest of
Your Life!* Dmitry Budker and Alexander O. Sushkov, Oxford University press. © Dmitry Budker, Alexander
O. Sushkov, Vasiliki Demas 2015, 2021. DOI: 10.1093/oso/9780198842361.003.0007

7.1   The number of elements in the periodic table

(**a**) Give a physical explanation why the *periodic table* has a finite number of elements, i.e., why atomic *nuclei* above a certain atomic number are unstable.

(**b**) A nucleus consists of protons and neutrons (nucleons) bound together by the *strong force*. Due to the "hard-core" close-range repulsion between nucleons, the nucleus can be approximated as a sphere of a constant density of nucleons. This is the *liquid drop model*. Suppose the nucleus contains $A = Z + N$ nucleons, where Z is the atomic number (number of protons), and N is the number of neutrons. Taking the size of a single nucleon to be $r_0 \approx 1\,\mathrm{fm}$, estimate the size of the entire nucleus.

(**c**) One of the simplest nuclei is the *deuteron*, which is a bound state of a proton and a neutron. The size of the deuteron is approximately $r_0 \approx 1\,\mathrm{fm}$. Use the uncertainty principle to estimate the depth of the nuclear potential V_s related to the strength of the strong force between nucleons.

(**d**) Using your answers to parts (b) and (c), estimate the atomic number of the heaviest stable element. Assume that the number of protons and neutrons in the nucleus is approximately equal: $Z \approx N \approx A/2$.

Solution

(a) The nucleons inside a nucleus are bound together by the strong force, which is approximately the same between proton-proton, neutron-neutron, and proton-neutron (this approximation is known as *isotopic invariance*). Protons carry positive electric charge, while neutrons are electrically neutral. Progressing down the periodic table towards the heavier elements (i.e., adding protons), the positive charge of the nucleus increases. Due to Coulomb repulsion, it costs more and more energy to add additional protons. The nucleon binding energy, however, remains approximately the same. When the Coulomb repulsion cost of adding a proton to the nucleus becomes larger than the nucleon binding energy, the nucleus becomes unstable, and can end its life via α-*decay*, β-*decay*, or *spontaneous fission*.

(b) Since the total number of nucleons is A, and their density is constant, the size of the entire nucleus is approximately

$$\boxed{R = r_0 A^{1/3}.} \tag{7.1}$$

(c) The wavefunctions of the nucleons in a deuteron are spatially confined on the length scale r_0. Therefore, from the uncertainty principle, the average momentum of each nucleon is $p \approx \hbar/r_0$. They thus have a kinetic energy $E \approx p^2/m_N$, where $m_N \approx 1\,\text{GeV}/c^2 \approx 2000 m_e$ is the nucleon mass, and m_e is electron mass. Assuming this kinetic energy is on the order of their interaction energy (see the three-dimensional case in Prob. 6.3), we estimate the strength of the strong interaction between nucleons:

$$\boxed{V_s \approx \frac{\hbar^2}{m_N r_0^2} = \frac{\hbar^2}{m_e a_B^2} \frac{m_e}{m_N} \frac{a_B^2}{r_0^2} \approx 30\,\text{MeV}.} \tag{7.2}$$

(d) Suppose we have a nucleus of charge $Z = A/2$, and radius $R = r_0 A^{1/3} = r_0(2Z)^{1/3}$. In order to add a proton, we have to overcome Coulomb repulsion

$$V_c = \frac{Ze^2}{R} \approx Z^{2/3} \frac{e^2}{a_B} \frac{a_B}{r_0} \approx 1\,\text{MeV} \times Z^{2/3}. \tag{7.3}$$

This Coulomb repulsion is equal to the nucleon strong interaction energy for

$$Z_{max} \approx (30\,\text{MeV}/1\,\text{MeV})^{3/2} \approx 160. \tag{7.4}$$

This is our estimate of the number of stable elements in nature. Nuclei with larger Z are unstable due to Coulomb repulsion between protons.

A more accurate calculation, considering fission of the nucleus into smaller nuclei, gives

$$\boxed{Z_{max} \approx 100.} \tag{7.5}$$

7.2   Neutron anatomy

The *nucleons*, the *neutron*, and the *proton*, are finite-sized particles, so it is perfectly legitimate to inquire about the distribution of the charge within them. A reasonable quantity that tells us how much charge there is at a radius r from the center of the nucleon is $4\pi r^2 \rho/e$, where the $4\pi r^2$ comes from the volume of a spherical shell of radius r and a fixed thickness dr, ρ is the electric-charge density, and e is the proton charge. These distributions for the proton and the neutron are sketched in Fig. 7.1. In the case of the neutron, we see that it qualitatively resembles an atom: there is a positive core, and a negative halo, while the whole system is neutral. In contrast to an atom, where the positive core (the nucleus) is several orders of magnitude more compact then the atom, the core is relatively larger for the neutron.

Give a qualitative explanation for the excess of negative charge at the periphery of the neutron.

Hint: Consider the *quark composition of the nucleons* and the concept of *virtual mesons*.

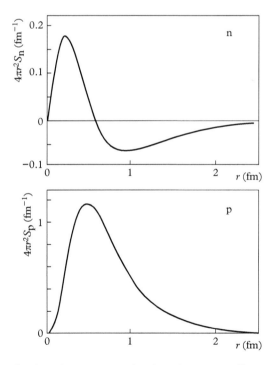

Fig. 7.1 Charge distributions in a neutron (top) and a proton (bottom). Like an atom, a neutron has a positive core and a negative "skin," though in vastly different spatial proportions. Please note the difference in the vertical scale.

Solution

The neutron is nominally a composite state of three quarks: ddu. The u quark has a charge of 2/3, and the d quark has a charge of -1/3, so the neutron is nicely neutral.

The u quark is light, and once in a while, a uū pair is virtually produced within the neutron (where ū is the antiparticle of u). We can write this like so:

$$ddu + u\bar{u} \;=\; duu + d\bar{u}. \tag{7.6}$$

Now, duu = uud is a proton, and dū with charge -1 is the π^- meson, which is much lighter than the proton. We can write this as a "reversible reaction":

$$n \rightleftarrows p + \pi^-. \tag{7.7}$$

Because the π^- meson is light, its radius is larger than that of the proton, and hence the positive core and the negative halo.

What about the proton? It can transform into a neutron and a positive meson:

$$p \rightleftarrows n + \pi^+, \tag{7.8}$$

which extends the positive charge density to larger radii, but does not change the fact that the charge density is positive everywhere.

7.3 Nonexistence of the dineutron

Explain why there is no *dineutron*—the bound state of two neutrons, even though the nucleus containing one proton and one neutron (the *deuteron*) is stable.

Hint: Assume that the *strong-interaction forces* are the same for a proton and a neutron (*isotopic invariance*), and make use of the fact that there is only one bound state for the proton-neutron system (the deuteron), and it is spin-one, mostly coming from the addition of the proton's and neutron's spins. (There is, in fact, a small admixture of the orbital angular momentum $L = 1$, but we can neglect it here.)

Solution

Isotopic invariance of the strong interactions tells us that the interaction potential in the deuteron and the dineutron should be the same. Coulomb forces are negligible in both cases due to the neutrality of the neutron. So, in the dineutron potential, we again have just one bound state, and as for the deuteron, it must be $S = 1$, $L = 0$ assuming isotopic invariance. However, this state is not accessible to the dineutron because the spin-triplet wavefunction with $L = 0$ is symmetric with respect to the interchange of the two neutrons, which is forbidden, for fermions, by the Pauli exclusion principle.

7.4   Deuterium fusion

(a) Write down possible fusion reactions involving two deuterium nuclei (^{2}H). Is ^{4}He formed in any of these reactions? Why?

(b) Roughly estimate the humankind's need for power.

(c) Using the facts that $\Delta E \approx 3\text{–}4\,\mathrm{MeV}$ of energy is released in each fusion process from part (a) and that ocean water contains ~ 150 parts per million (*ppm*) of deuterium (i.e., per one million hydrogen atoms, about 150 are deuterium), and making an optimistic assumption that someday we will learn to conduct highly efficient controlled thermonuclear reactions, estimate the volume of water we will need to use per year to supply the entire humankind with power.

Solution

(a) The possible fusion reactions are:

$$^2\text{H} + {}^2\text{H} \rightarrow {}^3\text{He} + \text{n} + \Delta E, \tag{7.9}$$

$$^2\text{II} + {}^2\text{II} \rightarrow {}^3\text{II} + \text{p} + \Delta E. \tag{7.10}$$

^{4}He is not formed because this would correspond to just one body in the final state, and it would be impossible to simultaneously satisfy the conservation of energy and momentum. One possibility would be the formation of an excited state of helium that would then decay via photon emission. However, this is suppressed because of the relatively long time scale of an electromagnetic decay.

(b) There are presently about eight billion people, and we need on the order of 1 kW of power per person averaged over time (a cooking plate consumes typically 1 kW, a bright light bulb—100 W, and a car—100 kW), so we need roughly 10^{13} W of power.

(c) Given the fact that a year is, to a good approximation, $\pi \times 10^7$ s, humankind needs about 3×10^{20} J in a year, which corresponds to about 2×10^{33} MeV. Given that 3–4 MeV is released per fusion, we need on the order of 10^{33} deuterons. Given the relative abundance of deuterium, this translates into $\sim 10^{37}$ hydrogen atoms in all. Now, 1 cm^3 of water weighs 1 g, and contains 1/18 moles of water, or $2 \times 6 \cdot 10^{23}/18 \approx 10^{23}$ hydrogen atoms. It follows that, to obtain the required amount of deuterium, we would need some 10^{14} cm^3 of water, or a cubic volume with about half a kilometer on the side (a small lake).

   This shows that, at least in principle, we have a practically unlimited energy supply right here on Earth.

7.5 Lifetime of the ground-state para-positronium

Positronium (Ps) is an atom that consists of an electron and its anti-particle a *positron*.

(a) What can you say about the energy-level structure of positronium?[1]

(b) Since both electron and positron are spin-1/2 particles, the total spin of positronium can be either $S = 0$ (*para-positronium*) or $S = 1$ (*ortho-positronium*). Which of these has lower ground-state energy, ortho- or para-positronium?

(c) Both ground-state ortho- and para-positronium can decay via annihilation of their constituents. However, it turns out that the lifetimes of these two states differ by three orders of magnitude. Shorter-lived para-positronium annihilates emitting two photons (Fig. 7.2), but this process is forbidden for ortho-positronium, so when it annihilates, three photons are emitted.

> There is more than one fundamental symmetry that forbids ortho-positronium to decay into two photons. One such prohibition comes from the so-called *Landau–Yang theorem* that says that, in a system of two photons, there is no state with total angular momentum one.

Here we will be concerned with two-photon annihilation of para-positronium. Give an order-of-magnitude estimate of para-positronium lifetime, which is determined by this process.

Hint: Use dimensional analysis and the fact that an amplitude of a process where a photon couples to an electron or a positron should be proportional to the coupling constant of the electromagnetic interaction, i.e., to the magnitude of the electron (positron) charge, e.

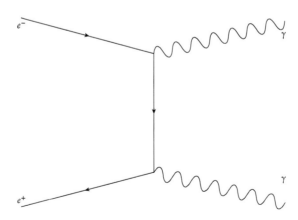

Fig. 7.2 A Feynman diagram representing positronium decay into two photons. Time is going left to right.

---

[1]Such an explicitly vague question is something that a student might encounter during an oral exam.

Solution

(a) The gross structure of the positronium energy spectrum is similar to that of hydrogen, but with an important difference that the relevant mass, the reduced mass of the two constituents, is almost a factor of two smaller for Ps than for H. Since all atomic-energy intervals are proportional to the reduced mass, all energies in Ps, including the ionization potential, are about one half of the corresponding energies in hydrogen.

The analogy between Ps and H for the case of fine and hyperfine structure is less direct: the two effects in Ps are of the same order because the magnetic moment of the "nucleus" is on the order of a Bohr magneton in Ps, and it is several orders of magnitude smaller in H.

(b) We can try to use an analogy with the hyperfine structure of the hydrogen ground state. Here, due to the *Fermi contact interaction* between the electron and the proton, the state with total angular momentum $F = 1$ lies above the $F = 0$ state by an energy interval corresponding to the famous *21-cm line*. In positronium, we may correspondingly expect that the triplet state (ortho-positronium) lies higher than the singlet state (para-positronium), and that the energy separation is 2–3 orders of magnitude ($\sim m_p/m_e$) larger than in H.

> While this is all true, it turns out that there is an additional interaction of the same order of magnitude that needs to be taken into account in Ps (and not in H!), and which is necessary to include to obtain agreement with experiment. This interaction is due to the *virtual* electron-positron annihilation and *pair production* processes. A detailed exposition of this topic can be found in (Series, 1988).

(c) We will ignore all numerical factors of order unity for the purpose of the estimate.

In order for the electron and the positron to annihilate, they should be "close" to each other. What sets the scale for the proximity length? Looking at the Feynman diagram for the annihilation process, we can think of annihilation as a two-step processes along the lines of what is shown in Fig. 7.2. In the first step, one of the two particles emits a photon; in the second step, the same particle meets with the other one, at which point they both disappear, and a photon is emitted. Note that the emission of the first photon violates the energy conservation, which is OK because of the uncertainty relations, but only for a short time. Since the energy of each of the photons is approximately equal to the rest energy of the electron (positron), we conclude that the relevant length and time scales are given here by the *Compton wavelength* and *Compton* time:

$$\lambda_c = \frac{\hbar}{m_e c}, \quad \tau_c = \lambda_c/c = \frac{\hbar}{m_e c^2}. \tag{7.11}$$

We will express the lifetime of the para-positronium as a product of the characteristic time τ_c and some dimensionless quantity.

Taking into account the hint, we are now ready to put it all together to estimate the para-positronium lifetime. The characteristic time has to be multiplied by the inverse fractional probability of the two particles to be found in a volume of a cubic Compton wavelength

$$\approx \left[\frac{\hbar/(m_e c)}{a_0} \right]^3 \qquad (7.12)$$

divided by e^4 (two photons are emitted, so the amplitude goes as e^2, and the probability i.e., the rate of the process goes as e^4; we divide because the lifetime is reciprocal to the rate). Since we are not allowed to have any quantities with dimensions in our formula apart from the Compton time, e^4 actually needs to be written as $[e^2/(\hbar c)]^2 = \alpha^2$. With this, we arrive at the following order-of-magnitude estimate:

$$\boxed{\tau_{Ps} \approx \frac{\hbar}{m_e c^2} \left[\frac{a_0}{\hbar/(m_e c)} \right]^3 \frac{1}{\alpha^2} = \frac{\hbar}{m_e c^2} \frac{1}{\alpha^5}.} \qquad (7.13)$$

As it frequently happens, despite of our blatant neglect of numerical constants, this result is only a factor of two shorter than the actual para-positronium lifetime of about 0.1 ns.

7.6 Schwinger fields

Our modern view of *vacuum* is that it is far from being empty space. In fact, vacuum is full of things, for example, spontaneously appearing and annihilating particle-antiparticle pairs.

In the presence of an electromagnetic field applied to vacuum that is stronger than some critical value called the *Schwinger field* in honor of one of the pioneers of *quantum electrodynamics (QED)*, virtual particles can acquire sufficient energy from the interaction with the field to create pairs of real particles.

(a) Estimate the value of the Schwinger field considering the effect of the field on a virtual electron-positron pair.

(b) The best way to produce strong fields is to spatially focus ultrashort laser-light pulses. What is the intensity of the light corresponding to the Schwinger field?

(c) What about the magnetic field?

Solution

(a) Production of a virtual particle of mass m requires energy on the order of mc^2 (where c is the light speed). In vacuum, this energy can appear for a *Compton time*

$$t_c = \frac{\hbar}{mc^2}, \tag{7.14}$$

where $\hbar$ is the reduced Planck constant. The range of distances over which a virtual particle travels during t_c is the reduced *Compton length*

$$\lambda_c = t_c \cdot c = \frac{\hbar}{mc^2}. \tag{7.15}$$

The energy acquired by a virtual electron in the presence of an electric field E can be estimated as

$$E \approx e\mathrm{E}\lambda_c = \frac{e\mathrm{E}\hbar}{mc}. \tag{7.16}$$

For the electric field equal to the Schwinger field ($\mathrm{E}=\mathrm{E}_c$), this energy is equal to mc^2, from which we arrive at

$$\boxed{\mathrm{E}_c = \frac{m^2 c^3}{e\hbar} \approx 10^{16}\,\mathrm{V/cm} \approx 4 \cdot 10^{13}\,\mathrm{esu.}} \tag{7.17}$$

(b) Light intensity is given by the *Poynting vector*

$$\frac{c}{4\pi}\mathrm{E}_c^2 \approx 4 \cdot 10^{36}\,\mathrm{esu} = \boxed{4 \cdot 10^{29}\,\mathrm{W/cm}^2.} \tag{7.18}$$

So far, the highest intensities achieved with lasers come a few orders of magnitude short of this value.

(c) In Gaussian units, electric field and magnetic induction are measured in the same units (gauss), so we can read the value of the Schwinger magnetic field B_c from Part (a). Another way of thinking about this is to consider the magnetic energy of a virtual electron in a magnetic field (B) and equate it to the rest mass of the electron for $\mathrm{B}=\mathrm{B}_c$:

$$E \approx 2\frac{e\hbar}{2mc}\mathrm{B}_c = mc^2, \tag{7.19}$$

yielding

$$\boxed{\mathrm{B}_c = \frac{m^2 c^3}{e\hbar} \approx 4 \cdot 10^{13}\,\mathrm{G.}} \tag{7.20}$$

The factor of 2 in Eq. (7.19) represents the electron-spin gyromagnetic ratio and $e\hbar/(2mc)$ is the Bohr magneton.

7.7 Cherenkov radiation

When a charged particle moves through a medium with a speed exceeding the phase velocity of light in this medium, it produces so-called *Cherenkov radiation*.

While a detailed microscopic description of the formation of Cherenkov radiation can be tricky, some properties of the radiation can often be described, to a good approximation, by remarkably simple formulas. For instance, the total energy lost by the charged particle per unit length is given by the *Frank–Tamm formula* (see Sec. 20–7 of Panofsky and Phillips 2005):

$$\frac{dE}{dL} = -Z^2\alpha \int_\omega \left(1 - \frac{1}{n^2(\omega)\beta^2}\right)\hbar\omega\,\frac{d\omega}{c}, \tag{7.21}$$

where Z is the charge of the particle in units of the *elementary charge* e, α is the *fine-structure constant*, ω is the frequency of the emitted light, $n(\omega)$ is the *refractive index* of the medium, $\beta = v/c$ is the speed of the particle normalized by the speed of light in vacuum, and the integration in Eq. (7.21) is carried out over the light frequencies for which the *phase velocity* of light in the medium c/n is less than the speed of the particle βc, i.e., for $n(\omega)\beta > 1$.

(a) Suppose an *ultrarelativistic* ($\beta \approx 1$) charged particle propagates through a low-density atomic vapor, for instance, an evacuated vapor cell containing rubidium or cesium atoms at room temperature.

Qualitatively describe the spectrum of the Cherenkov radiation.

(b) The Cherenkov radiation is often described as forming a cone with the charged particle at the cone's apex. Equivalently, we can say that light is emitted at an angle θ_C with respect to the direction of the propagation of the particle.

Derive the expression for $\theta_C(\omega)$ in terms of $n(\omega)$ and β and re-express Eq. (7.21) in terms of this angle.

(c) An example where Cherenkov radiation can be seen by eye is the so-called pool-type *nuclear reactors*, where the core of the reactor is at the bottom of a water reservioir, the pool. Relativistic electrons originating from *fission reactions* produce beautiful blue light in the water (there are many pictures available on internet).

Why does the Cherenkov radiation appear blue in this case?

Solution

(a) For a dilute vapor, refractive index is close to unity at all frequencies, except for those where the refractive index is resonantly enhanced at the low-frequency side of the strong resonance transitions from the ground state (Fig. 7.3). Thus the spectrum of the Cherenkov radiation consists of a series of sharp lines corresponding to these transitions.

(b) The standard (and still elegant) derivation of the angle of the Cherenkov radiation assumes that each point on the linear trajectory of the particle becomes a secondary source of radiation whose wavefronts are on the surface of the sphere expanding with the phase velocity of light in the medium c/n. Considering two such points separated by a distance L, the time delay between when the particle passes through these points is $L/(\beta c)$, and we find that the wavefront (where the radiation is in-phase) of the Cherenkov radiation propagates at an angle $\theta_C = \arccos\left[(c/n)/(\beta c)\right] = \arccos\left[(n\beta)^{-1}\right]$. With this, Eq. (7.21) becomes

$$\boxed{\frac{dE}{dL} = -Z^2\alpha \int_\omega \sin^2\theta_C(\omega)\hbar\omega\frac{d\omega}{c}.}$$

(7.22)

(c) Questions regarding *color perception* could be tricky as they may involve physiology in addition to physics.

Let us begin by noticing that the refractive index of water changes only slightly over the visible range ($n \approx 1.33$ at 700 nm, and ≈ 1.34 at 400 nm). While the increase

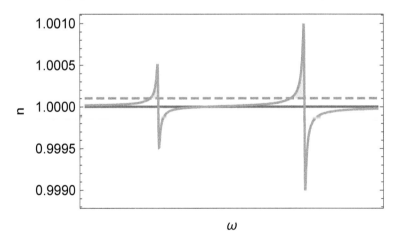

Fig. 7.3 Cherenkov radiation in a dilute atomic vapor is emitted at frequencies close to strong resonant transitions from the ground state, where the refractive index of the medium is resonantly enhanced (indicated by the filled regions on the plot). The vertical span on this plot depends on the experimental conditions such as the temperature of the vapor cell. The dashed horizontal line indicates the value of $1/\beta$.

of the refractive index towards the short-wave side of the spectrum favors production of Cherenkov radiation there, we neglect this effect.

Assuming n constant, Eq. (7.21) tells us that the spectral density of the radiation increases linearly with frequency. In order to infer how such radiation is perceived by a human eye, we compare this spectrum with that of the *blackbody radiation*.

If we observe an object as it is heated, it first appears dark red, and then becomes brighter as it is heated further, while the perceived color changes towards lighter colors. Once the object is a few thousand kelvin hot, we perceive it as white. The spectral density of the blackbody radiation at 5000 K is sketched in Fig. 7.4. We see from the figure that the blackbody spectral density drops towards shorter wavelengths over the visible range (indicated on the figure).

The opposite slope of the Cherenkov radiation and the (apparently white) blackbody spectrum make it plausible that the perceived Cherenkov spectrum is shifted towards the higher-frequency part of the spectrum, and thus appears blue.

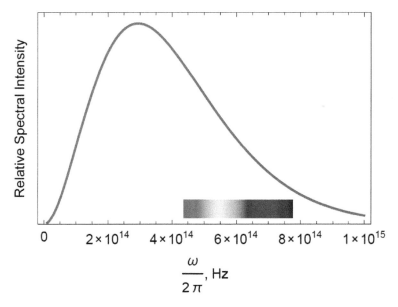

Fig. 7.4 Spectrum of the blackbody radiation at 5000 K. The filled rectangle shows the visible range of the spectrum. The Cherenkov radiation has an opposite slope over the visible range and is perceived as blue.

7.8   Neutron optics

In analogy with light interacting with transparent medium, the interaction of slow (cold; Fig. 7.5) neutrons with sufficiently large *de Broglie wavelength* with nonabsorbing materials can be described using the concept of *refractive index*. In both cases, the refractive index can be defined as the ratio of the magnitudes of the wave vector in the medium (K) and that in vacuum (k):

$$n = \frac{K}{k}. \tag{7.23}$$

It turns out that for most (but not all!) materials, the potential energy for a neutron in the medium is higher than in vacuum, and neutron refractive index is less than unity $(n < 1)$, which, according to the fundamental relation

$$K = \frac{2\pi}{\lambda}, \tag{7.24}$$

means that the neutron de Broglie wavelength increases when the neutron enters from vacuum into such material. The latter is also true for a photon entering a medium with $n < 1$.

(a) What can we say about the change in the photon and neutron velocity upon entering the material? (For simplicity, assume that the neutron enters the material perpendicular to the surface; for the case of the photon, in this problem it is sufficient to discuss the phase velocity.) Explain the difference.

(b) What is the neutron *phase velocity* and *group velocity* in free space?

Solution

(a) In the photon case, the phase velocity goes from the speed of light in vacuum c to c/n. (There is no problem with phase velocity exceeding c for $n < 1$ in terms of *special relativity* as no signals can be sent at the phase velocity.)

For the case of the neutron, we can determine the velocity in the following way. The momentum of the neutron in the medium is $p = \hbar K = n\hbar k$, so its velocity is

$$v = \frac{p}{m} = \frac{n\hbar k}{m}, \tag{7.25}$$

where m is the mass of the neutron. This is slower that the neutron's speed in vacuum $\hbar k/m$.

The "opposite" behavior of the photon and neutron velocities is a consequence of the difference in their *dispersion relations*. For instance, in vacuum, the photon speed is independent of its momentum, but the neutron speed is proportional to its momentum.

(b) *Phase velocity* is defined in terms of the frequency ω and the wavevector k of a wave as

$$v_{ph} = \frac{\omega}{k}, \tag{7.26}$$

while *group velocity* is defined as

$$v_{gr} = \frac{\partial \omega}{\partial k}. \tag{7.27}$$

For a neutron in vacuum, it is natural to set the frequency defined up to an arbitrary offset as

$$\omega = \frac{p^2}{2m\hbar} = \frac{\hbar k^2}{2m}, \tag{7.28}$$

which yields $\boxed{v_{ph} = p/(2m)}$ and $\boxed{v_{gr} = p/m.}$ The latter recovers the classical expression, while the former is one half the classical value.

Fig. 7.5 A "cold" neutron.

8
Solid-State Physics

News from the South Pole: room
temperature superconductivity has
been discovered!

Physics on Your Feet: Berkeley Graduate Exam Questions: or Ninety Minutes of Shame but a PhD for the Rest of Your Life! Dmitry Budker and Alexander O. Sushkov, Oxford University press. © Dmitry Budker, Alexander O. Sushkov, Vasiliki Demas 2015, 2021. DOI: 10.1093/oso/9780198842361.003.0008

8.1   Transistors

(**a**) Explain the operation of a *field-effect transistor* (FET), where the current between emitter and collector is controlled by voltage applied to base.

(**b**) Explain the operation of a *bipolar-junction transistor* (BJT), where the current between emitter and collector is controlled by the current flowing through base.

Solution

(a) A FET has three terminals: a source, a drain, and a gate. The body (or substrate) of the FET defines the voltage reference, biasing the device; it is often connected to the source terminal, see Fig. 8.1. The flow of the charge carriers (electrons or holes) from the source to the drain is controlled by the voltage applied between the gate and the source. The current that flows through the gate terminal is usually small, thus, a FET is a voltage-controlled device, and has a large input resistance. There are several types of FETs, depending on the sign of the charge carriers (n-channel and p-channel), and the behavior of the conductive channel between the source and the drain (depletion-mode and enhancement-mode). For example, in an n-channel, depletion-mode FET, the width of the conductive channel through which electrons flow from source to drain is controlled with the gate voltage. A negative gate voltage expands the depletion region, narrowing this channel and increasing its resistance, and, at the pinch-off gate voltage, the resistance becomes so large that the FET is turned off, like a switch.

The most common types of FETs are junction field-effect transistor (JFET) in which a reverse biased p–n junction separates the gate from the body, and metal-oxide-semiconductor field-effect transistor (MOSFET) in which an insulating layer, usually SiO_2, separates the gate from the body. FETs are commonly used as amplifiers or switches.

(b) A BJT has three terminals: an emitter, a base, and a collector. These are connected to three differently-doped regions of semiconductor, denoted as pnp or npn, in contact with each other. The current between the emitter and the collector is controlled by a much smaller base-emitter current, thus, a BJT is a current-controlled device.

Let us focus on the simple model of operation of an npn *transistor*. The principle of operation relies on the diffusion of carriers inside the base region of the device. The n-type emitter region has a high doping concentration, and the thermally-excited electrons can diffuse across the depletion region between the emitter and the base. Once inside the base region, a small fraction of these electrons recombines with one of the holes there (this is the base current), but the base region is thin enough so that most of the electrons diffuse to the base-collector junction. There they see a strong electric field, created by the reverse-biased depletion region, that accelerates them into the collector region. These electrons constitute the emitter-collector current. Changes in the small base current control the width of the emitter-base depletion region, and thus the number of electrons that can diffuse across this region into the base.

The bipolar transistor was invented in 1947 by Bardeen, Brattain, and Shockley. This discovery was a key milestone in the birth of the semiconductor electronics industry.

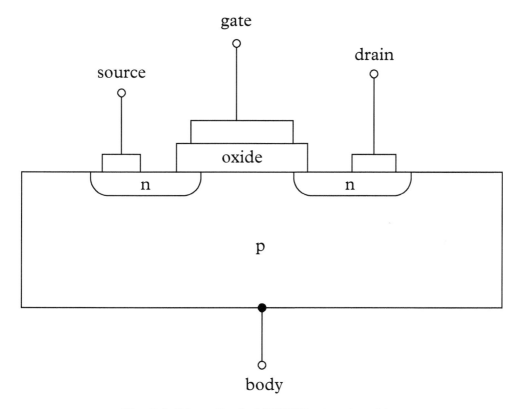

Fig. 8.1 Schematic of a MOSFET n-type transistor.

BIPOLAR TRANSISTOR

8.2 Magnetic domains

Consider a particle of an isotropic ferromagnetic material of characteristic linear dimensions a. In a *ferromagnet*, atomic spins tend to line up in a common direction. However, it turns out that particles that are larger than some characteristic size a^* tend to split into oppositely oriented *domains*.

From the most general consideration, give a rough estimate of the critical size a^*.

Solution

We begin by recalling the reasons why the adjacent spins in a ferromagnet tend to orient in the same direction (see, for example, Ch. 12 of Kittel 2007 or Ch. 37 of Feynman et al. 1989). This has to do with a quantum mechanical phenomenon called *exchange interaction* that leads to an "energy cost" of J for two adjacent spins to be antiparallel, which is usually some fraction of atomic energy (we will assume $J \approx 0.01\,Ry$ for an estimate; note that the exchange energy also sets the scale for the ferromagnetic-transition temperature a.k.a. the *Curie temperature* of the ferromagnet).

Now, consider our ferromagnetic particle with all spins lined up in the same direction [Fig. 8.2 (a,b)]. We can mentally separate the particle in two parts. Let us first do this as shown with the dashed line in Fig. 8.2 (a). The two resulting magnetic particles are in a stable state because the energy of two magnets is minimized when they are "on top of each other." However, the two magnets resulting from splitting the particle as in Fig. 8.2 (b) are in a state of high energy, which is minimized if one of the sides reverses its magnetization as shown in Fig. 8.2 (c).

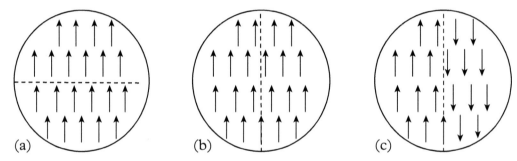

(a)   (b)   (c)

Fig. 8.2 A uniformly magnetized ferromagnetic particle (a) may be mentally split into two. Since magnets that are "on top of each other" actually tend to line up this way, there is no reason for the two parts in (a) not to remain magnetized as they are. On the other hand, if we mentally split the particle as shown in (b), the two resulting parts are in a state with higher dipole-dipole-interaction energy compared to the situation shown in (c). The flip from (b) to (c) occurs when the dipole-dipole-energy difference exceeds the "energy cost" of having spins on either side of the domain wall being antiparallel.

Such a splitting of one magnetic domain into two minimizes the dipole-dipole interaction, but "costs" energy because the spins on the two sides of the *domain wall* are now antiparallel. Let us equate the energy gain and energy cost.

Neglecting numerical factors of order unity, we write the dipole-dipole energy as

$$\frac{\left(\mu_B na^3\right)^2}{a^3} \approx \frac{\mu_B^2 a^3}{a_L^6}. \tag{8.1}$$

Here, we assumed that the magnetic moment per atom is on the order of the Bohr magneton (μ_B; for iron, the value is close to $2\mu_B$ per atom), and that the number density is $n \approx 1/a_L^3$, where a_L is the *lattice constant*, i.e., a characteristic distance between atoms in the material. Typically, $a_L \approx 6a_0$, where a_0 is the Bohr radius.

The number of atoms in the domain wall can be estimated as $(a/a_L)^2$, and so the energy cost of the spins being antiparallel on either side of the wall is something like

$$J\left(\frac{a}{a_L}\right)^2. \tag{8.2}$$

Equating the two energies for $a = a^*$, we arrive at:

$$\frac{\mu_B^2 a^3}{a_L^6} \approx J\left(\frac{a}{a_L}\right)^2 \Rightarrow \boxed{a^* \approx J\frac{a_L^4}{\mu_B^2} \approx J\frac{a_L^4}{\alpha^2(ea_0)^2} \approx a_0 \frac{J}{\alpha^2 Ry}\left(\frac{a_L}{a_0}\right)^4.} \tag{8.3}$$

Here, we made use of the facts that $\mu_B = (\alpha/2)ea_0$, where $\alpha \approx 1/137$ is the *fine-structure constant*, e is the elementary charge, and that $Ry = e^2/(2a_0)$. The numerical value of the previous expression is $a^* \approx 100$ μm, which is to be compared with typical values of several microns for real materials, not that bad given the crudeness of our approximations.

A more accurate estimate (which is more than what we want to do here) can be done with a more sophisticated and realistic model. In the previous we assumed that the domain wall thickness is a single lattice constant, i.e., the spins on either side of the wall simply flip from "up" to "down." In reality, the exchange energy can be lowered if this transition happens gradually, so that the angle between adjacent spins is not $180°$, but $180°/N$, the domain wall then has thickness Na_L. In addition to the exchange and magnetic interactions, magnetic anisotropy plays a role in the physics of ferromagnetic materials. This interaction favors the alignment of magnetization along certain crystalline directions. It is caused by *spin-orbit interaction*, and is thus suppressed by a factor of $\alpha^2 Z^2$ (Z being the atomic number) with respect to the exchange interaction; the magnitude of the anisotropy energy is on the order of 10^{-4} eV per atom for transition metals. It is the balance between the exchange interaction and the magnetic anisotropy that sets the width (N) of the domain wall. For iron, $N \approx 300$. The revised theory would take into account the finite wall thickness by revising the calculation of the wall energy.

But wait! In view of this "derivation," how is it possible to have magnets that are of macroscopic (for example, centimeter) size? Why do not such magnets demagnetize by splitting into a bunch of differently oriented domains? In fact, macroscopic magnets do consist of such domains. Under applied magnetic field, these domains rotate or change size (by domain wall movement), changing the magnetization of the material. However, domain magnetization can become "stuck" either due to crystallite (grain) size in magnetic ceramics, or due to *domain-wall pinning* by impurities and crystal defects, sometimes purposely introduced. This leads to *magnetic hysteresis* and *coercivity*. Thus, even when the applied magnetic field is removed, the material can retain its magnetization.

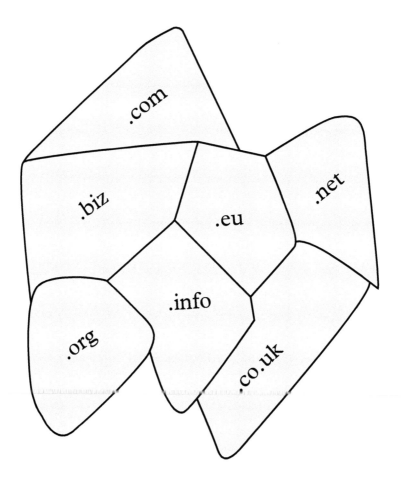

8.3 Damaging diamond

One of many known *defects in diamond crystals* is a *vacancy*, i.e., a missing carbon atom in the diamond lattice. A vacancy can be formed if one imparts a kinetic energy of 40 eV (or more) to a carbon atom in the lattice, thus "kicking" the atom out of the lattice.

(a) Estimate the minimum energy of an *electron beam* irradiating the diamond that is needed to produce vacancies. Is it important to include relativistic effects in this estimate?

(b) Suppose we need to produce uniformly distributed vacancies in a diamond plate that is 1 mm thick using a beam of electrons. What energy electrons should be used?

(c) In practice, researchers tend to avoid using electrons with energies in excess of 15 MeV or so for this purpose. Can you think of a reason why?

Solution

(a) The maximum energy is transferred from an electron to a carbon nucleus in a head-on collision, when the electron is retro-reflected. At 40 eV, the motion of the carbon nucleus is deeply non-relativistic, but we cannot a-priori assume this for the electron. If the electron momentum magnitude is p, the magnitude of the transferred momentum in a head-on collision is $\approx 2p$, and the kinetic energy of the carbon nucleus (and atom) is $\approx 4p^2/(2M)$, where $M \approx 12$ GeV/c^2 is the mass of the carbon atom. Equating this to 40 eV, we get $p \approx 0.5$ MeV/c, so the relativistic effects are, in fact, non-negligible.

Evaluating the kinetic energy, K, of the electron from

$$K = E - mc^2, \tag{8.4}$$

$$E^2 = p^2 c^2 + m^2 c^4, \tag{8.5}$$

where E is the total energy of the electron, and m is its mass, we find that the required kinetic energy of the electrons is about 200 keV.

(b) When an electron propagates through a medium, it loses energy to the medium's electrons (*ionization loss*). This is described by the *Bethe formula* that can be written in a simplified form:

$$\frac{dE}{dx} \approx -\frac{2 \text{ MeV}}{\beta^2} \frac{\rho \cdot \text{cm}^2}{\text{g}}. \tag{8.6}$$

Here $\beta = v/c$ and v is the electron's velocity. For relativistic electrons $\beta \approx 1$, the rate of energy loss is independent of energy. For diamond, $\rho \approx 3.5$ g/cm^3, and the loss rate is about 7 MeV/cm. Once the electron loses enough energy so it is no longer ultra-relativistic, β becomes smaller than unity, and the rate of energy loss progressively increases. For our case of a 1 mm thick diamond plate, electrons of a few MeV and higher in energy are needed to uniformly damage the sample.

(c) At these energies, electrons are capable of *activation* of the nuclei in the sample, making the sample *radioactive*. This is undesirable in most cases.

8.4 Nearest-neighbor defect in a crystal

Consider a crystalline solid with a concentration of *point defects* of n (defects per unit volume) that are randomly distributed in the crystal with a characteristic distance between adjacent defects ($n^{-1/3}$) much larger than the size of the crystal's unit cell.

What is the distribution of the distances between defects and their nearest-neighbor defects? This is relevant in evaluating perturbations, e.g., the dipole-dipole interaction, between the defects. If the perturbations are sufficiently short-range, they may be dominated by the nearest neighbors.

Solution

Let us first consider the probability $P_1(r)$ that there are no other defects in a sphere of radius r from a given defect. The boundary condition is $P_1(0) = 1$.

Knowing $P_1(r)$ for a certain radius $r > 0$, we determine

$$P_1(r + dr) = P_1(r) \cdot \left(1 - n \cdot 4\pi r^2 dr\right), \tag{8.7}$$

where the probability of having no defects in a sphere of radius $r + dr$ is written as a product of such probabilities for a sphere of radius r and a thin spherical shell of radius r and thickness dr. Equation (8.7) leads to a differential equation that is solved to yield:

$$P_1(r) = e^{\frac{-4\pi n}{3} r^3}. \tag{8.8}$$

The probability $P(r)$ of finding the closest neighbor in a shell of radius r and thickness dr is given by the probability of having no defects in a sphere of radius r, i.e., $P_1(r)$, times the probability of finding a defect in the shell. This leads to the sought-for distribution:

$$\boxed{dP(r) = e^{\frac{-4\pi n}{3} r^3} 4\pi n r^2 dr.} \tag{8.9}$$

8.5   Fermi velocity in a metal

What is the order of magnitude of the *Fermi velocity* of electrons in a metal? Please express the answer in terms of fundamental constants.[1]

$$-\frac{\hbar}{i}\frac{\partial}{\partial t} = \frac{p^2}{2m} - \frac{Ze^2}{r}$$

$$\alpha = \frac{\hbar^2}{ec}$$

Artist's impression of the great Enrico Fermi that is based on a famous photograph. Can you see what is wrong with this picture?

---

[1] This question was inspired by a discussion with Prof. Victor Flambaum of the University of New South Wales.

Solution

A good initial approximation to describe *electron gas* in a metal is to assume that the *conduction-band electrons* are free and that each of the atoms in the metal contributes on the order of one electron to the conduction band (Kittel, 2007).

The *density of states* of the electrons including the two possible polarizations for a given state is (see, for example, Prob. 3.6) is

$$\text{Density of quantum states} = \frac{2 \cdot 4\pi p^2 dp}{(2\pi\hbar)^3}, \tag{8.10}$$

where $p = mv$ is the electron's momentum. The density of electrons with momentum not exceeding the Fermi momentum p_F is

$$\frac{2 \cdot 4\pi p_F^3}{3 \cdot (2\pi\hbar)^3}, \tag{8.11}$$

which should equal the density of the conduction-band electrons, in turn equal to the number density of the metal. Assuming, for the purpose of an order-of-magnitude estimate that the distance between the atoms in the metal is on the order of the *Bohr radius* a_0, we have

$$\frac{m^3 v_F^3}{3\pi^2\hbar^3} = \frac{1}{a_0^3}, \tag{8.12}$$

where v_F is the *Fermi velocity*. From this, it follows that $\boxed{v_F \approx \alpha c}$, where $\alpha = e^2/(\hbar c) \approx 1/137$ is the *fine structure constant*. This shows that electrons in a metal are nonrelativistic and that the Fermi velocity is comparable to the characteristic speed of valence electrons in atoms.

8.6   Superfluid transition of helium

In this problem we calculate the superfluid transition temperature for helium, neglecting interactions between helium atoms.

(**a**) Neutral ^{4}He atoms consist of a nucleus with two protons and two netrons, as well as two electrons. Are these atoms bosons or fermions?

(**b**) What is the relationship between energy and momentum (dispersion relation) for helium atoms in a non-interacting gas?

(**c**) Briefly explain the phenomenon of Bose condensation. Explain what happens to occupation numbers of quantum energy states of an ensemble of atoms undergoing Bose condensation. The *chemical potential* is the amount of energy it takes to add one ^{4}He atom to the ensemble; explain what happens to the chemical potential in the condensate phase.

(**d**) Calculate the Bose condensation temperature T_c for ^{4}He in terms of helium atom mass m and number density n. In your calculation you may use the following integral:

$$\int_0^\infty \frac{\sqrt{x}dx}{e^x - 1} \approx 2.3.$$

(**e**) Calculate the thermal de Broglie wavelength of the ^{4}He atoms at temperature T_c and compare it to the spacing between the atoms.

(**f**) Approximately evaluate T_c for helium atoms.

Solution

(a) Since there is an even number of particles of each type, the total spin of a ^{4}He atom is an integer, which makes it a boson. In fact, protons, neutrons, and electrons all have spin $1/2$ and the spin angular momenta add up to the total angular momentum zero in the ground state.

(b) Since we are neglecting interactions between helium atoms, only kinetic energy is relevant, with the usual energy-momentum relationship

$$E = p^2/2m,$$

where m is the mass of a helium atom.

(c) Bose–Einstein condensation describes a transition to the state of matter with a macroscopic occupation of the ground state. The chemical potential μ of a Bose–Einstein condensate of non-interacting helium atoms is zero: adding a particle to the ensemble costs no energy, since it can be added directly to the ground state $E = 0$.

(d) The Bose–Einstein distribution gives the occupation number for a quantum state with energy $E = p^2/2m$ at temperature T, and for ensemble chemical potential μ:

$$\eta = \frac{1}{e^{(E-\mu)/k_B T} - 1}.$$

We take the condensation temperature $T = T_c$ to be the temperature at which the chemical potential of the ensemble is $\mu = 0$:

$$n = \int_0^\infty \frac{\eta \, d^3 p}{(2\pi\hbar)^3} = \int_0^\infty \frac{d^3 p}{(2\pi\hbar)^3} \frac{1}{e^{p^2/2mk_B T_c} - 1},$$

where n is the atom number density. Substituting $x = p^2/2mk_B T_c$ we can simplify the integral:

$$n = \frac{4\pi\sqrt{2}}{(2\pi\hbar)^3} (mk_B T_c)^{3/2} \int_0^\infty \frac{\sqrt{x} dx}{e^x - 1},$$

and using the given numerical value for the integral, we obtain:

$$\boxed{T_c \approx 3.3 \frac{\hbar^2 n^{2/3}}{mk_B}.}$$

(e) The de Broglie wavelength of a particle with momentum p is defined as $\lambda = h/p$. For the ^{4}He atoms at $T = T_c$, the characteristic thermal momentum is given by $p = \sqrt{2mk_B T_c}$, and, using the answer to part (d), the de Broglie wavelength is

$$\lambda = h/\sqrt{2mk_B T_c} \approx \sqrt{6} n^{-1/3},$$

i.e., the de Broglie wavelength is equal to the inter-particle spacing up to a factor of order unity. This is the equivalent condition for Bose–Einstein condensation: it occurs at the temperature at which the de Broglie wavelength becomes comparable to the spacing between non-interacting particles.

(**f**) We use the density of liquid helium: $n \approx 10^{22} \, \text{cm}^{-3}$ and the mass of a helium atom: $m \approx 10^{-23}$ g, together with values of fundamental constants: $\hbar \approx 10^{-34} \, \text{J} \cdot \text{s}$, $k_B \approx 10^{-23}$ J/K to estimate the T_c for helium atoms:

$$T_c \approx 3 \, \text{K},$$

which is close to the experimentally observed temperature of the liquid helium super-fluid transition of 2.17 K.

8.7 Frustrated spins

The concept of *frustration* arises in condensed matter physics, for example in the context of spins located on a triangular lattice, when the spins "can not decide" which way they would like to point, or, more precisely, when the quantum ground state of the system is degenerate. In this problem, we explore a simple toy model for such a system.

Consider a system of three $s = 1/2$ spins located at the vertices of an equilateral triangle and interacting with each other through the Heisenberg interaction with strength J. Such a system might be realized on a triangular lattice. The Hamiltonian is given by:

$$H = J(s_1 s_2 + s_2 s_3 + s_3 s_1) - \mu H s_1, \tag{8.13}$$

where $s_i = \pm 1/2$ $(i = 1, 2, 3)$ are projections on magnetic field H, $J, \mu, H > 0$, and, for simplicity, only one of the spins couples to the magnetic field. The spin system is coupled to a heat bath at temperature T.

(a) List all the states of the system together with their energies. What is the energy and degeneracy of the ground state of the system?

(b) What is the mean energy of the system for $T \to 0$?

(c) What is the mean energy of the system for $T \to \infty$?

(d) What is the entropy of the system for $T \to 0$?

(e) What is the entropy of the system for $T \to \infty$?

(f) Write the partition function for this system at temperature T.

(g) Write the mean energy of the system as a function of temperature.

Solution

(a) Here are the states of the system and their energies:

$$
\begin{aligned}
(+ + -) &\to -J/4 - \mu H/2,\\
(+ - -) &\to -J/4 - \mu H/2,\\
(+ - +) &\to -J/4 - \mu H/2,\\
(- + +) &\to -J/4 + \mu H/2,\\
(- + -) &\to -J/4 + \mu H/2,\\
(- - +) &\to -J/4 + \mu H/2,\\
(+ + +) &\to 3J/4 - \mu H/2,\\
(- - -) &\to 3J/4 + \mu H/2.
\end{aligned}
\tag{8.14}
$$

(b) At $T \to 0$, the system goes to its ground state with energy $\bar{E} = -J/4 - \mu H/2$ (3-fold degenerate).

(c) For $T \to \infty$, the system samples all energy states equally, therefore we have to take the mean of all energies in Eq. (8.14), resulting in $\bar{E} = 0$.

(d) We use Boltzmann's equation for entropy: $S = k_B \ln W$, where W is the number of microstates that the system can occupy. For $T \to 0$, the system is in the ground state, which is threefold degenerate, thus $W = 3$ and $S = k_B \ln 3$.

(e) For $T \to \infty$, system is equally likely to be in all 8 energy states, thus $W = 8$ and $S = k_B \ln 8$.

(f) The partition function is defined as the sum over all microstates j of the system:

$$
\begin{aligned}
Z = \sum_j e^{-E_j/k_B T} =\\
= 3e^{-(-J/4 - \mu H/2)/k_B T} + 3e^{-(-J/4 + \mu H/2)/k_B T} +\\
+ e^{-(3J/4 - \mu H/2)/k_B T} + e^{-(3J/4 + \mu H/2)/k_B T}.
\end{aligned}
\tag{8.15}
$$

(g) The mean energy can also be calculated as the sum over all microstates j of the system:

$$
\begin{aligned}
\bar{E} = \frac{1}{Z} \sum_j E_j e^{-E_j/k_B T} =\\
= \frac{1}{Z}\Big[-3(J/4 + \mu H/2)e^{(J/4 + \mu H/2)/k_B T} - 3(J/4 - \mu H/2)e^{(J/4 - \mu H/2)/k_B T} +\\
+ (3J/4 - \mu H/2)e^{-(3J/4 - \mu H/2)/k_B T} + (3J/4 + \mu H/2)e^{-(3J/4 + \mu H/2)/k_B T}\Big].
\end{aligned}
\tag{8.16}
$$

Appendix A
Maxwell's Equations and Electromagnetic Field Boundary Conditions

Two physicists are condemned to death, but are given one last wish each. The first one says: "I have so many ideas, that I would like to give a seminar for my colleague as my last wish." The second says: "I wish to die before he gives the talk!"

Here we compile the equations that may be of use for solving electromagnetism problems in this book. We use Gaussian units.

The four Maxwell's equations are:

$$\boldsymbol{\nabla} \cdot \mathbf{E} = 4\pi\rho, \tag{A.1}$$

$$\boldsymbol{\nabla} \cdot \mathbf{B} = 0, \tag{A.2}$$

$$\boldsymbol{\nabla} \times \mathbf{E} = -\frac{1}{c}\frac{\partial \mathbf{B}}{\partial t}, \tag{A.3}$$

$$\boldsymbol{\nabla} \times \mathbf{B} = \frac{1}{c}\left(4\pi\mathbf{j} + \frac{\partial \mathbf{E}}{\partial t}\right). \tag{A.4}$$

The electric and magnetic fields in presence of polarizable materials are defined as follows:

$$\mathbf{D} = \mathbf{E} + 4\pi\mathbf{P}, \tag{A.5}$$

$$\mathbf{H} = \mathbf{B} - 4\pi\mathbf{M}. \tag{A.6}$$

For linear materials

$$\mathbf{D} = \epsilon\mathbf{E}, \tag{A.7}$$

$$\mathbf{H} = \mathbf{B}/\mu. \tag{A.8}$$

The continuity equations reads

$$\frac{\partial \rho}{\partial t} + \boldsymbol{\nabla} \cdot \mathbf{j} = 0. \tag{A.9}$$

The Lorentz force on a charge q in electromagnetic field is

$$\mathbf{F} = q \left(\mathbf{E} + \frac{\mathbf{v}}{c} \times \mathbf{B} \right). \tag{A.10}$$

We also quote the interface boundary conditions for electric and magnetic fields at a boundary between two media, with $\mathbf{n}_{12}$ being the normal vector from medium 1 to medium 2.

- Tangential component of $\mathbf{E}$ is continuous across the surface:

$$\mathbf{n}_{12} \times (\mathbf{E}_2 - \mathbf{E}_1) = 0. \tag{A.11}$$

- Normal component of $\mathbf{D}$ is continuous if there is no charge on the interface, otherwise the discontinuity is given by the surface charge density:

$$\mathbf{n}_{12} \cdot (\mathbf{D}_2 - \mathbf{D}_1) = 4\pi\sigma. \tag{A.12}$$

- Normal component of $\mathbf{B}$ is continuous across the surface:

$$\mathbf{n}_{12} \cdot (\mathbf{B}_2 - \mathbf{B}_1) = 0. \tag{A.13}$$

- Tangential component of $\mathbf{H}$ is continuous if there is no current on the interface, otherwise the discontinuity is given by the surface current density:

$$\mathbf{n}_{12} \times (\mathbf{H}_2 - \mathbf{H}_1) = 4\pi\boldsymbol{\kappa}_s. \tag{A.14}$$

Appendix B
Symbols and Useful Constants

In Oregon (where there is no sales tax), a student buys an ice-cream for herself and her boyfriend, paying $17.39. She then realizes that the ice-cream vendor is a crook. How did she figure this out? The answer depends on whether she is a mathematics student ($17.39 does not divide into two), or a physics student ($17 is too much for the ice-cream!).

B.1 Symbols

Here, we compile the list of symbols used in the book.

Table B.1: Some notations used throughout this book.

Symbol	Meaning
E	energy
K	kinetic energy
U	potential energy
H	Hamiltonian or height
F	force or total angular momentum
P	pressure or power
T	temperature
Q	quality factor
m	mass
ρ	density
v	velocity
t	time
$\mathbf{E}$, E	electric field vector and magnitude
$\mathbf{B}$, B	magnetic induction vector and magnitude
$\mathbf{D}$	electric displacement vector
$\mathbf{H}$	magnetic field vector
$\mathbf{M}$	magnetization
$\mathbf{S}$	Poynting vector
Φ	magnetic flux
I	current or moment of inertia
j	current density
d	electric dipole moment
σ	surface charge density or conductivity
ε	dielectric permittivity
μ	magnetic dipole moment or material permeability
λ	wavelength
ω	angular frequency

B.2   Useful constants

The following is a table of some constants and their values that may be useful to the reader.

Symbol	Meaning	Value
m, m_e	electron mass	9.1085×10^{-28} g
		0.511 MeV/c^2
m_p	Planck mass	10^{19} GeV/c^2
e	electron charge magnitude	4.8029×10^{-10} esu
h	Planck's constant	6.6252×10^{-27} erg·s
$\hbar = h/(2\pi)$		1.0544×10^{-27} erg·s
$\alpha = e^2/(\hbar c)$	fine structure constant	$1/137.036$
$a_0 = \hbar^2/(me^2)$	Bohr radius	5.292×10^{-9} cm
$\mu_B = e\hbar/(2mc)$	Bohr magneton	0.93×10^{-20} erg/G
		1.40 MHz/G
$\mu_N = e\hbar/(2m_p c)$	nuclear magneton	5.06×10^{-24} erg/G
		762 Hz/G
$Ry = me^4/(2\hbar^2)/(hc)$	Rydberg constant	$109,737$ cm^{-1}
k_B	Boltzmann constant	1.38066×10^{-16} erg/K
		8.61735×10^{-5} eV/K
N_A	Avogadro number	6×10^{23}
g	acceleration of free fall on the earth surface	980 cm/s^2

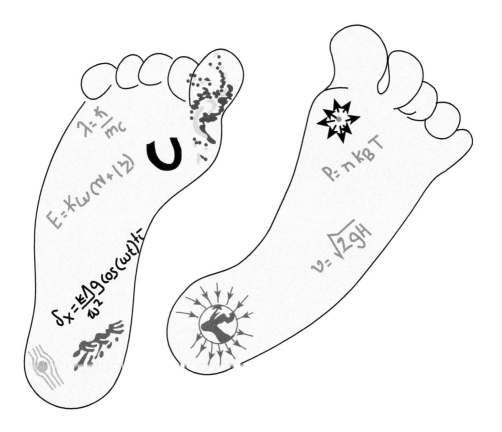

References

Auzinsh, M., Budker, D., and Rochester, S.M. (2010). *Optically Polarized Atoms: Understanding Light-Atom Interactions*. New York: Oxford University Press.

Abbott, B. P. et al., (LIGO collaboration) (2017). "GW170814: A Three-Detector Observation of Gravitational Waves from a Binary Black Hole Coalescence." *Physical Review Letters*, **119**: 141101.

Abbott, B. P. et al., (LIGO collaboration) (2017). "The basic physics of the binary black hole merger GW150914." *Annalen der Physik*, **529**: 1600209.

Bai, Gui-ru, Guo, Guang-can, Lim, Yung-kuo, The Physics Coaching Class University of Science and Technology, Zhongguo (1991). *Problems and Solutions on Optics: Major American Universities Ph.D. Qualifying Questions and Solutions*. Singapore and New Jersey: World Scientific.

Band, Yehuda and Avishai, Yshai (2012). *Quantum Mechanics With Applications to Nanotechnology and Information Science*. Oxford: Academic Press.

Battesti, R, Beard, J., Böser, S., Bruyant, N., Budker, D., Crooker, S. A., Daw, E. J., Flambaum, V. V., Inada, T., Irastorza, I. G., Karbstein, F., Kim, D. L., Kozlov, M. G., Melhem, Z., Phipps, A., Pugnat, P., Rikken, G., Rizzo, C., Schott, M., Semertzidis, Y. K., ten Kate, H. H. J., Zavattini, G., (2018). "High magnetic fields for fundamental physics." *Physics Reports*, **765-766**: 1–39.

Brown, R. Hanbury (1974). *The Intensity Interferometer: Its Application to Astronomy*. London, New York: Taylor and Francis; Halsted Press.

Budker, Dmitry, Kimball, Derek F., and DeMille, David (2008). *Atomic Physics. An Exploration Through Problems and Solutions* (2nd edn). Oxford: Oxford University Press.

Cahn, Sidney B., Mahan, Gerald D., and Nadgorny, Boris E. (1994). *A Guide to Physics Problems*. The Language of Science. New York: Plenum Press.

Chen, Min (1974). *University of California, Berkeley, Physics Problems, with Solutions*. Englewood Cliffs, N.J: Prentice-Hall.

Chou, C. W., Hume, D. B., Rosenband, T., and Wineland, D. J. (2010). "Optical Clocks and Relativity." *Science*, **329**: 1630–1633.

Cronin, Jeremiah A., Greenberg, David F., and Telegdi, Val (1967). *University of Chicago Graduate Problems in Physics, with Solutions*. Addison-Wesley Series in Advanced Physics. Reading, Massachusetts: Addison-Wesley.

Dall, R. G., Hodgman, S. S., Manning, A. G., Johnsson, M. T., Baldwin, K. G. H., and Truscott, A. G. (2011). "Observation of Atomic Speckle and Hanbury Brown-Twiss Correlations in Guided Matter Waves." *Nature Communications*, **2**, 291.

Darrigol, Olivier (2009). *Worlds of Flow: A History of Hydrodynamics from the Bernoullis to Prandtl*. New York: Oxford University Press.

Einstein, A. (1918). "Über Gravitationswellen." *Sitzungsberichte der Königlich Preußischen Akademie der Wissenschaften (Berlin), Seite 154-167*.

Feynman, Richard P., Leighton, Robert B., and Sands, Matthew L. (1989). *The Feynman Lectures on Physics*. Redwood City, California: Addison-Wesley.

Galitski, Victor, Karnakov, Boris, Kogan, Vladimir, and Galitski, Jr., Victor (2013). *Exploring Quantum Mechanics*. Oxford: Oxford University Press.

Heald, Mark A. and Marion, Jerry B. (1995). *Classical Electromagnetic Radiation*. Philadelphia: Saunders College Publishing.

Helmholz, A. Carl (2004). *History of the Physics Department, University of California, Berkeley 1950–1968*. University of California, Berkeley.

Herzberg, Gerhard (1989). *Molecular Spectra and Molecular Structure, Volume 1: Spectra of Diatomic Molecules*. Krieger Florida: Publishing Company.

Huber, M. E. and Cabrera, B. and Taber, M. A. and Gardner, R. D. (1990). "Limit on the flux of cosmic-ray magnetic monopoles from operation of an eight-loop superconducting detector." *Physical Review Letters*, **64**: 835–838.

Jackson, John David (1998). *Classical Electrodynamics, 3rd ed.* New York: John Wiley & Sons.

Khriplovich, I. B. (1991). *Parity Nonconservation in Atomic Phenomena*. New York: Gordon and Breach.

Kittel (2007). *Introduction To Solid State Physics, 7th ed.* New York: John Wiley & Sons Inc.

Landau, Lev Davidovich and Litshitz, E. M. (1991). *Quantum Mechanics: Non-Relativistic Theory, 3rd ed.* Oxford: Butterworth-Heinemann.

Landau, Lev Davidovich and Lifshitz, E. M. (1999). *Mechanics*. Oxford: Butterworth-Heinemann.

Lim, Yung-kuo (1998). *Problems and Solutions on Quantum Mechanics*. Major American Universities Ph.D. Qualifying Questions and Solutions. World Scientific, Singapore; River Edge, NJ.

Lim, Yung-kuo (2000). *Problems and Solutions on Atomic, Nuclear and Particle Physics*. Major American Universities Ph.D. Qualifying Questions and Solutions. World Scientific, Singapore; River Edge, NJ.

Loudon, Rodney (2000). *The Quantum Theory of Light,* 3rd edn. Oxford Science Publications. Oxford; New York: Oxford University Press.

Mandel, L. and Wolf, E. (1995). *Optical Coherence and Quantum Optics.* Cambridge: Cambridge University Press.

Migdal, A. B. and Krainov, V. P. (1969). *Approximation Methods In Quantum Mechanics.* New York, W.A. Benjamin.

Milonni, Peter W. (1994). *The Quantum Vacuum. An Introduction to Quantum Electrodynamics.* San Diego, California: Academic Press.

Nagourney, Warren (2014). *Quantum Electronics for Atomic Physics*, 2nd edn. New York: Oxford University Press.

Panofsky, Wolfgang Kurt Hermann and Phillips, Melba (2005). *Classical Electricity and Magnetism.* New York: Dover Publications.

Peebles, P. J. E. (1993). *Principles of Physical Cosmology.* Princeton Series in Physics. New Jersey: Princeton University Press.

Povh, B. and Rosina, M. (2005). *Scattering and Structures: Essentials and Analogies in Quantum Physics.* Berlin, New York: Springer.

Series, G. W. (1988). *The Spectrum of Atomic Hydrogen Advances: A Collection of Progress Reports by Experts.* Singapore: World Scientific.

Siegman, A. E. (1986). *Lasers.* Sausalito, California: University Science Books.

Siu, M.S. and Budker, D. (1999). "Solidification Pipes: From Solder Pots to Igneous Rocks." *arXiv:physics/9911057.*

Zhang, Jia-lü, Zhou, You-yuan, Zhang, Shi-ling, and Lim, Yung-kuo (1995). *Problems and Solutions on Solid State Physics, Relativity and Miscellaneous Topics.* Major American universities Ph.D. Qualifying Questions and Solutions. Singapore; River Edge: World Scientific: NJ.

Index